Animal Tracks
of Southern California

ANIMAL
TRACKS
OF
SOUTHERN
CALIFORNIA

IAN SHELDON

LONE
PINE

© 1998 by Lone Pine Publishing
First printed in 1998 10 9 8 7 6 5 4 3 2 1
Printed in Canada

THE PUBLISHER: LONE PINE PUBLISHING

1901 Raymond Ave. SW, Suite C	206, 10426–81 Avenue	202A, 1110 Seymour St.
Renton, WA 98055	Edmonton, AB T6E 1X5	Vancouver, BC V6B 3N3
USA	Canada	Canada

Canadian Cataloguing in Publication Data
Sheldon, Ian, date–
    Animal tracks of Southern California

    Includes bibliographical references and index.
    ISBN 1-55105-105-2

    1. Animal tracks–California, Southern–Identification. I. Title.
QL768.S627 1998   591.47'9   C97-911048-3

Senior Editor: Nancy Foulds
Editor: Volker Bodegom
Production Manager: David Dodge
Design, layout and production: Volker Bodegom, Gregory Brown
Map: Volker Bodegom
Technical review: Donald L. Pattie
Animal illustrations: Gary Ross, Horst Krause, Kindrie Grove
Track illustrations: Ian Sheldon
Cover illustration: Gary Ross
Scanning: Elite Lithographers Ltd., Edmonton, Alberta, Canada
Printing: Quality Color Press Inc., Edmonton, Alberta, Canada

The publisher gratefully acknowledges the support of the Department
of Canadian Heritage.

CONTENTS

INTRODUCTION

If you have ever spent time with an experienced tracker, or perhaps a veteran hunter, then you know just how much there is to learn about the subject of tracking and just how exciting the challenge of tracking animals can be. Maybe you think that tracking is no fun, because all you get to see is the animal's prints. What about the animal itself–isn't that much more exciting? Well, for most of us who don't spend a great deal of time in the beautiful wilderness of southern California, the chances of seeing the graceful Pronghorn at full speed or spotting the secretive Bobcat are slim. The closest we may ever get to some animals will be through their tracks, and these can inspire a very intimate experience. Remember, you are following in the footsteps of the unseen–animals that are in pursuit of prey, or per- haps being pursued as prey.

This book offers an introduction to the complex world of tracking animals. Sometimes tracking is easy. At other times it is an incredible challenge that leaves you wonder- ing just what animal left those unusual tracks. Take this book into the field with you as it can provide some help with the first steps to identification. Prints and tracks are this book's focus; you will learn to recognize subtle differ- ences for both. There are, of course, many additional signs to consider, such as scat and food caches, all of which help you to understand the animal that you are tracking.

Remember, it takes many years to become an expert tracker. Tracking is one of those skills that grows with you as you acquire new knowledge in new situations. Most importantly, you will have an intimate experience with nature. You will learn the secrets of the seldom seen. The more you discover, the more you will want to know, and by developing a good understanding of tracking, you will gain an excellent appreciation of the intricacies and delights of our marvelous natural world.

How to Use this Book

Most importantly, take this book into the field with you! Relying on your memory is not an adequate way to identify tracks. Track identification has to be done in the field, or with detailed sketches and notes that you can take home. Much of the process of identification is circumstantial, so you will have much more success when standing beside the track.

This book is laid out so as to be easy to use. There is a quick reference to the tracks of all the animals illustrated in this book beginning on p. 152. This appendix is a fast way to familiarize yourself with certain tracks and the content of the book, and it guides you to the more informative descriptions of each animal and track.

Each animal's description is illustrated with the appropriate footprints and the styles of tracks that it usually leaves. While these illustrations are not exhaustive, they do show the tracks or groups of prints that you will most likely see. Where there are differences in orientation, left prints are illustrated. You will find a list of dimensions for the tracks, giving the general range, but there will always be extremes, just as there are with people who have unusual small or large feet. Under the category 'Size' (of animal), the 'greater-than' sign (>) is used when the size difference between the sexes is pronounced.

If you think that you may have identified a track, check the 'Similar Species' section for that animal. This section is designed to help you confirm your conclusions by pointing out other animals that leave similar tracks and showing you ways to distinguish among them.

As you read this book, you will notice an abundance of words such as 'often,' 'mostly' and 'usually.' Unfortunately, tracking will never be an exact science; we cannot expect animals to conform to our expectations, so be prepared for the unpredictable.

Tips on Tracking

As you flip through this guide, you will notice clear, well-formed prints. Do not be deceived! It is a rare track that will ever show so clearly. For a good, clear print, the perfect conditions are slightly wet, shallow snow that isn't melting, or slightly soft mud that isn't actually wet. Needless to say, these conditions can be rare–most often you will be dealing with incomplete or faint prints, where you cannot even really be sure of the number of toes.

Should you find yourself looking at a clear print, then the job of identification is much easier. There are a number of key features to look for: Measure the length and width of the print, count the number of toes, check for claw marks and note how far away they are from the body of the print,

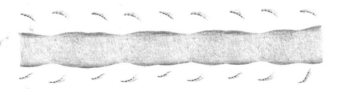

and look for a heel. Keep in mind other, more subtle features, such as the spacing between the toes and whether they are parallel or not, and whether fur on the sole of the foot has made the print less clear.

When you are faced with the challenge of identifying an unclear print—or even if you think that you have made a successful identification from one print alone—look beyond the single footprint and search out others. Do not rely on the dimensions of one print alone, but collect measurements from several prints to get an average impression. Even the prints within one track can show a lot of variation.

Try to determine which is the fore print and which is the hind, and remember that many animals are built very differently from humans, having larger forefeet than hind feet. Sometimes the prints will overlap, or they can be directly on top of one another in a direct register. For some animals, the fore and hind prints are pretty much the same.

Check out the pattern that the prints make together in the track, and follow the trail for as many paces as is necessary for you to become familiar with the pattern. Patterns are very important, and can be the distinguishing feature between different animals with otherwise similar tracks.

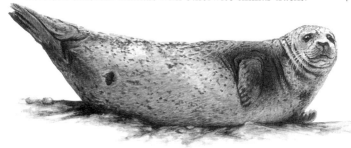

Follow the trail for some distance, because it may give you some vital clues. For example, the trail may lead you to a tree, indicating that the animal is a climber, or it may lead down into a burrow. This part of tracking can be the most rewarding, since you are following the life of the animal as it hunts, runs, walks, jumps, feeds or tries to escape a predator.

Take into consideration the habitat. Sometimes very similar species can be distinguished only by their habitats—one might be found on the riverbank, while another might be encountered only in the dense forest.

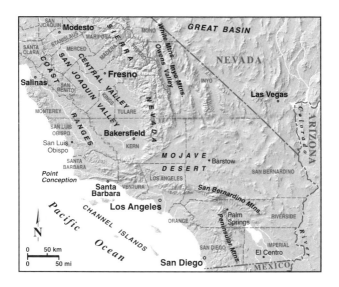

Think about your geographical location, too. Some animals have a limited range, perhaps only in remote parts of California—or they may be confined to the desert or to the mountains. This consideration can rule out some species and help you with your identification.

Remember that every animal will at some point leave a print or track that looks just like the print or track of a completely different animal!

Lastly, keep in mind that if you track quietly, you might catch up with the maker of the prints.

Terms and Measurements

Some of the terms used in tracking can be rather confusing, and they often depend on personal interpretation. For example, what comes to your mind if you see the word 'hopping'? Perhaps you see a person hopping about on one leg, or perhaps you see a rabbit hopping through the countryside. Clearly, one person's perception of motion can be very different from another's. Some useful terms are explained below, to clarify what is meant in this book, and where appropriate, how the measurements given fit in with each term.

The following terms are sometimes used loosely and interchangeably—for example, a rabbit might be described as 'a hopper' and a squirrel as 'a bounder,' yet both leave the same pattern of prints in the same sequence.

Bounding: Can be used interchangeably with 'hopping' or 'jumping'; often used for patterns that cover short distances; hind prints usually registering ahead of fore prints.

Galloping: Used for the motion made by animals with four even legs, such as dogs, moving at speed; hind prints registering ahead of fore prints.

Hopping: Similar to bounding; usually indicated by tight clusters of prints; fore prints set between and behind the hind prints.

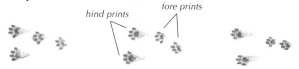

Running: Like galloping, but often applied generally to animals moving at speed.

Stotting (applies to Mule Deer only): Describes the action of taking off from the ground and landing on all four feet at once, in pogo-stick fashion.

Trotting: Faster than walking, slower than running.

Other Tracking Terms

Alternating track: A left-right sequence, as made by humans walking; often a double register for four-legged animals, which are described as 'diagonal walkers.'

double register   *direct register*

Dewclaws: Two small, toe-like structures set above and behind the main foot of most hoofed animals.

dew claws

dragline

Direct register: The hind print falls directly on the fore print.

Double register: The hind print overlaps the fore print only slightly or falls beside it, so that both can be seen at least in part.

Dragline: The line left in snow or mud by a foot or the tail dragging over the surface.

Gallop group: A cluster of four prints made at a gallop, usually with hind prints registering ahead of fore prints.

Height: Taken at the animal's shoulder.

Length: The animal's body length from head to rump, not including the tail, unless otherwise indicated.

Lope: A collection of four prints made at a fast pace, usually falling roughly in a line.

Print: Fore and hind prints are treated individually; print dimensions are length (including claws—maximum values may represent occasional heel register for some animals) and width; together the prints make up a track.

Register: To leave a mark—said about a foot, claw or other part of an animal's body.

Retractable: Describes claws that can be pulled in to keep them sharp, as with the cat family; these claws do not register in the prints.

Sitzmark: The mark left on the ground by animal falling or jumping from a tree.

Straddle: The total width of the track, all prints considered.

Stride: For consistency among different animals, the stride is taken as the distance from the center of one print (or print group) to the center of the next one. Other books may use the term 'pace.'

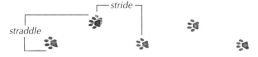

Track: A pattern left by a series of prints.

Trail: Often used to describe a track at length; think of it as the path of the animal.

MAMMALS

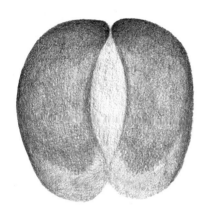

Fore and Hind Prints
Length: 4.0–6.0 in (10–15 cm)
Width: 4.0–6.0 in (10–15 cm)

Straddle
10–21 in (25–53 cm)

Stride
Walking: 14–32 in (36–81 cm)

Size (bull>cow)
Height: 5.0–6.0 ft (1.5–1.8 m)
Length: 10–13 ft (3.0–4.0 m)

Weight
Male: 800–2000 lb (360–910 kg)
Female: 700–1100 lb (320–500 kg)

walking

BISON (Buffalo)
Bos bison

The magnificent Bison, frequently called 'buffalo,' once roamed North America in numbers estimated at 70 million. As few as 1500 remained after wholesale slaughter of the Bison in the nineteenth century. Today, as many as 100,000 Bison roam selected protected areas and public and private ranches, resulting in a very scattered distribution, notably on Santa Catalina Island. Do not be fooled by a Bison's calm exterior: a bull Bison can inflict serious injury!

In a Bison's alternating walking pattern, the slightly smaller hind print usually registers on or near the fore. On firm ground, only the outer edge of the hoof may register, but in soft mud, the whole print registers–perhaps the dewclaws too. The abundant 'pies' may be mistaken for the dung of domestic cattle. Additional signs of Bison include rubbing posts or trees with tufts of distinctive brown hair hanging from them and the large pits in which they wallow.

Similar Species: Domestic Cattle (*Bos* spp.) prints are similar. Horse (p. 30) prints can resemble Bison prints on firm surfaces.

Fore and Hind Prints
Length: 3.2–5.0 in (8.1–13 cm)
Width: 2.5–4.5 in (6.4–11 cm)

Straddle
7.0–12 in (18–30 cm)

Stride
Walking: 16–34 in (41–86 cm)
Galloping: 3.3–7.8 ft (1.0–2.4 m)
Group length: to 6.3 ft (1.9 m)

Size (stag>hind)
Height: 4.0–5.0 ft (1.2–1.5 m)
Length: 6.5–10 ft (2.0–3.0 m)

Weight
500–1000 lb (230–450 kg)

gallop print *walking*

ELK
(Wapiti)
Cervus elaphus

The Elk, largest of the California deer, once had a wide range but it is now much reduced. Its scattered distribution, notably in the Owens Valley, is largely the result of successful reintroductions. Female Elk and young are often seen in social herds as they feed in forest openings and lush meadows; they move into the valleys when winter sets in. Stags, who prefer to go solo, are easily recognized by their magnificent racks of antlers and haunting bugling.

Elk usually leave a neat alternating track with large, rounded prints, often in well-worn winter paths. The hind print will sometimes double register slightly ahead of the fore print. In deeper snow, or if an Elk gallops (with its toes spread wide), the dewclaws may register. A good place to look for Elk tracks is in the soft mud by summer ponds, where Elk like to drink and sometimes splash around.

Similar Species: The smaller Mule Deer's (p. 22) similar walking track has generally shorter, narrower prints.

Fore and Hind Prints
Length: 2.0–3.3 in (5.1–8.4 cm)
Width: 1.6–2.5 in (4.1–6.4 cm)

Straddle
5.0–10 in (13–25 cm)

Stride
Walking: 10–24 in (25–61 cm)
Stotting: 9.0–19 ft (2.7–5.8 m)

Size (buck>doe)
Height: 3.0–3.5 ft (91–110 cm)
Length: 4.0–6.5 ft (1.2–2.0 m)

Weight
100–450 lb (45–200 kg)

walking   *stotting group*

MULE DEER
(Black-tailed Deer)
Odocoileus hemionus

The widespread Mule Deer is frequently seen in meadows, open woodlands and arid plains. In winter, it moves down from higher terrain to warmer, south-facing slopes and sagebrush flats. It is absent from the San Joaquin Valley and the deserts. In the Coast Ranges, where it is slightly smaller than elsewhere, it is commonly called 'Black-tailed Deer.'

The Mule Deer has a neat, alternating walking track, with the hind print registering on the fore. In winter, this deer prefers to stay in small groups and frequently uses the same well-worn trail. Mule Deer prints are heart-shaped and sharply pointed. In deeper snow or when a deer is moving quickly in mud, its prints show dewclaws, which are closer to the toes on the fore print. At speed, these deer jump with all their feet leaving and striking the ground at once: 'stotting.' Stotting tracks show how the toes spread to distribute the weight and give better footing.

Similar Species: A Pronghorn Antelope (p. 24) print has a wider base. Elk (p. 20) have longer, wider prints.

23

Fore and Hind Prints
Length: 3.3 in (8.4 cm)
Width: 2.5 in (6.4 cm)
Straddle
3.5–9.0 in (8.9–23 cm)
Stride
Walking:
    8.0–19 in (20–48 cm)
Galloping:
    14 ft (4.3 m) or more
Size (buck>doe)
Height: 3.0 ft (91 cm)
Length: 3.8–4.9 ft (1.2–1.5 m)
Weight
75–130 lb (34–59 kg)

walking

gallop group

PRONGHORN ANTELOPE
Antilocapra americana

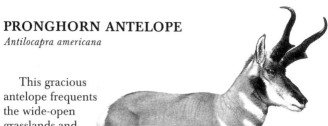

This gracious antelope frequents the wide-open grasslands and sagebrush plains at higher altitudes in the eastern regions, especially in the dry terrain east of the Sierra Nevada. Pronghorns gather in groups of up to a dozen animals in summer and as many as one hundred in winter; they prefer to feed in areas where the snow has blown away. The Pronghorn, one of the fastest animals in North America, runs for fun, easily attaining constant speeds of 40 mph (65 km/h) and short bursts of up to 60 mph (95 km/h).

A Pronghorn print has a pointed tip and a broad base. This animal does not have dewclaws. The hind prints usually register directly on top of the fore prints, making a tidy alternating track. Pronghorns tend to drag their feet in snow. During their frequent gallops, their toe tips spread wide. The faster the antelope moves, the greater the distance between gallop groups.

Similar Species: Mule Deer (p. 22) and Elk (p. 20), with dewclaws and narrower toe bases, respectively, make stotting tracks and have shorter strides between gallop groups.

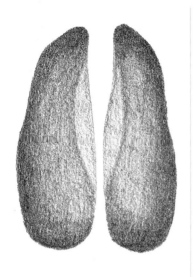

Fore and Hind Prints
Length: 2.5–3.5 in (6.4–8.9 cm)
Width: 1.8–2.5 in (4.6–6.4 cm)

Straddle
6.0–12 in (15–30 cm)

Stride
Walking: 14–24 in (36–61 cm)

Size (ram>ewe)
Height: 2.5–3.5 ft (76–110 cm)
Length: 4.0–6.5 ft (1.2–2.0 m)

Weight
75–270 lb (34–120 kg)

walking

BIGHORN SHEEP
(Mountain Sheep, Desert Bighorn)
Ovis canadensis

In late fall, the loud crack of two majestic rams head-butting one another can be heard for a great distance. To watch the rut is an awe-inspiring experience. Found in the barren lands of desert country in summer, this sheep prefers high, arid terrain and mountain slopes, but moves into valleys during winter. It tends to avoid forested areas.

The print is quite squarish in shape and pointed towards the front. The outer edge of the hoof is hard, while the inner part is soft, giving the sheep a good grip on tricky terrain. Find one track, and you will likely come across several, because this sheep is gregarious and wants to be with others of its kind. The neat alternating walking pattern is a direct or double register of the hind over the fore. When this sheep runs, its toes spread wide. Bighorn tracks may lead you to sheep beds—hollows dug into the snow that are used many times and often have a large accumulation of droppings.

Similar Species: A Domestic Sheep (*Ovis* spp.) print is similar. A Mule Deer (p. 22) print is more heart-shaped.

**Fore and Hind Prints
(with dewclaws)**
Length: 2.5–3.0 in (6.4–7.6 cm)
Width: 2.3 in (5.8 cm)
Straddle
5.0–6.0 in (13–15 cm)
Stride
Trotting: 16–20 in (41–51 cm)
Size (male>female)
Height: 3.0 ft (91 cm)
Length: 4.3–6.0 ft (1.3–1.8 m)
Weight
77–440 lb (35–200 kg)

trotting

FERAL PIG (Wild Pig, Wild Boar)
Sus scrofa

Descended from introduced European animals, the Feral Pig has mixed with escapees of domestic stock. Populations of these sturdy beasts of the dense undergrowth can be found scattered at lower elevations through much of the state, sometimes roaming about in groups of as many as fifty. Armed with strong tusks, they can be quite threatening.

A Feral Pig's print shows two prominent, widely spaced toe marks, and usually, except on firm surfaces, a clear, pointed dewclaw mark off to each side. The hind print is slightly smaller than the fore print. Feral Pigs are keen foragers, so their tracks can often be numerous, especially when they travel in a group. They usually trot, with a typical track showing a double register of hind over fore print, in the alternating pattern typically made by four-legged animals. Other signs are wallows and diggings.

Similar Species: Mule Deer (p. 22) tracks are similar, but with longer stride, pointier prints, a narrower gap between the toes and dewclaws to the rear rather than to the side. The Domestic Pig (also *Sus scrofa*) has a wider straddle and less neat tracks that often form two separate lines.

29

Fore Print
(hind print is slightly smaller)
Length: 4.5–6.0 in (11–15 cm)
Width: 4.5–5.5 in (11–14 cm)

Stride
Walking: 17–27 in (43–69 cm)

Size
Height: to 6.0 ft (1.8 m)

Weight
to 1500 lb (680 kg)

walking

HORSE
Equus caballus

This popular animal has unmistakable prints; it deserves mention because back-country use of the Horse means that you can expect its tracks to show up almost anywhere.

Unlike any other animals discussed in this book, the Horse has only one huge toe, which leaves an oval print. A distinctive feature is the 'frog' or V-shaped mark at the base of the print. When a Horse is shod, the horseshoe shows up clearly as a firm wall at the outside of the print. Not all horses will be shod, so don't expect to see this outer wall on every horse track.

A typical, leisurely horse track is an alternating walk with hind prints registering on or behind the slightly larger fore prints. Horses are capable of a range of speeds—up to a full gallop—but most recreational horseback riders prefer to walk their horses and soak up the mountain views!

Similar Species: Mules have smaller prints and they are rarely shod.

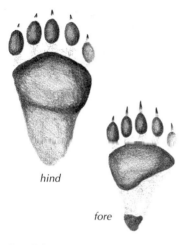

hind

fore

Fore Print
Length without heel:
    4.0–6.3 in (10–16 cm)
Width: 3.8–5.5 in (9.7–14 cm)

Hind Print
Length: 6.0–7.0 in (15–18 cm)
Width: 3.5–5.5 in (8.9–14 cm)

Straddle
9.0–15 in (23–38 cm)

Stride
Walking: 17–23 in (43–58 cm)

Size (male>female)
Height: 3.0–3.5 ft (91–110 cm)
Length: 5.0–6.0 ft (1.5–1.8 m)

Weight
200–600 lb (91–270 kg)

walking
(slow)

BLACK BEAR
Ursus americanus

The Black Bear is widespread in forested, mountainous areas, especially in the Sierra Nevada. Black Bears sleep deeply in the winter, so don't expect to encounter their tracks in the colder months. Finding fresh bear tracks can be a thrill, but take care, as the bear may be just around the corner. Never underestimate the potential power of a surprised bear!

A Black Bear's prints are about the length of a human's, but wider and with claw marks. The small inner toe rarely registers. The forefoot's small heel pad often shows, and the hind foot has a big heel. The bear's slow walk results in a slightly pigeon-toed double register of the hind print over the fore. At a faster pace, the hind oversteps the fore. When a bear runs, the two hind feet register in front of the forefeet in an extended cluster. Along well-worn bear paths, look for 'digs'–patches of dug-up earth–and 'bear trees' whose scratched bark shows that these bears climb.

Similar Species: Grizzly Bear (*Ursus arctos*) prints are similar, but this massive bear has been driven out of the state.

fore

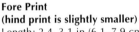

hind

Fore Print
(hind print is slightly smaller)
Length: 2.4–3.1 in (6.1–7.9 cm)
Width: 1.6–2.4 in (4.1–6.1 cm)
Straddle
4.0–7.0 in (10–18 cm)
Stride
Walking: 8.0–16 in (20–41 cm)
Gallop: 2.5–10 ft (76–300 cm)
Size
(female is slightly smaller)
Height: 23–26 in (58–66 cm)
Length: 32–40 in (81–100 cm)
Weight
20–50 lb (9.1–23 kg)

walking   *gallop group*

COYOTE (Brush Wolf, Prairie Wolf)
Canis latrans

This widespread and adaptable canine is especially abundant in the Great Basin. It prefers open areas or woodland, where it hunts rodents and larger prey, either on its own, with a mate or in a family pack. A Coyote also occasionally develops an interesting cooperative relationship with a Badger (p. 66), so you might find their tracks together where they have been digging for ground squirrels (p. 98).

The oval fore prints are slightly larger than the hind prints. Note the difference between the fore heel pad and the hind heel pad, which rarely registers clearly. Claw marks are usually evident only for the two center toes. A Coyote's tail hangs down, leaving a dragline in deep snow. Coyotes typically walk or trot in an alternating pattern—the walk has a wider straddle. When a Coyote gallops, the hind feet fall ahead of the forefeet; the faster it goes, the straighter the gallop group. A Coyote's trail is often direct, as if it knew exactly where it was going.

Similar Species: Domestic Dog (*Canis familiaris*) prints are not so oval and spread more, and a dog's trail is erratic and confused. Red Fox (*Vulpes vulpes*) prints are usually smaller.

35

fore

hind

Fore and Hind Prints
Length: 1.1–1.8 in (2.8–4.6 cm)
Width: 1.1–1.5 in (2.8–3.8 cm)

Straddle
2.0–4.0 in (5.1–10 cm)

Stride
Walking/Trotting:
    7.0–10 in (18–25 cm)

Size
Height: 12 in (30 cm)
Length with tail:
    24–31 in (61–79 cm)

Weight
3.0–6.0 lb (1.4–2.7 kg)

trotting

KIT FOX
(Swift Fox)
Vulpes velox

This small, shy fox hides away, on the plains and in desert regions in much of southern California–for example, at the southern end of the San Joaquin Valley and in the Mojave Desert. Secretive and solitary, its nocturnal activity makes it a rare sight, but evidence of its tracks can give it away. Its preference for sandy areas means that its tracks will seldom be very clear, as sand will often fall back into the print. If you find unclear prints, pay attention to general track characteristics, such as whether the prints are in a typical dog-family trotting pattern. The track's dimensions may help distinguish which species made them.

Similar Species: The Island Fox (*Urocyon littoralis*), found only in the Channel Islands, has similar prints. A Gray Fox's (p. 38) track is very similar, but slightly larger. The Red Fox (*Vulpes vulpes*), confined to the Sierra Nevada, has different heel pads (with a bar across them); in general, its prints are larger and less clear (because of thick fur), its stride longer and its straddle narrower. A print of the Domestic Cat (*Felis catus*) or the Bobcat (p. 42) will lack claw marks and have a larger, less symmetrical heel pad.

fore

hind

Fore Print
(hind print slightly smaller)
Length: 1.3–2.1 in (3.3–5.3 cm)
Width: 1.1–1.5 in (2.8–3.8 cm)

Straddle
2.0–4.0 in (5.1–10 cm)

Stride
Walking/Trotting: 7.0–12 in (18–30 cm)

Size
Height: 14 in (36 cm)
Length: 21–29 in (53–74 cm)

Weight
7.0–15 lb (3.2–6.8 kg)

walking

GRAY FOX

Urocyon cinereoargenteus

This small, shy fox is widespread but it especially prefers woodlands and chaparral country. This fox is the only one that climbs trees, which it does either for safety or to forage.

The larger fore print registers better than the hind print, on which the long, semi-retractable claws do not always show. The heel pads are often unclear–they sometimes show up just as small, round dots. This fox has a neat, alternating walking track. When it trots, its prints fall in pairs, with the front print set diagonally behind the rear print, and it has a gallop group like a Coyote's (p. 34).

Similar Species: The Island Fox (*Urocyon littoralis*) of the Channel Islands has smaller prints. The smaller Kit Fox (p. 36), usually with smaller prints, prefers desert plains and the south of the San Joaquin Valley. The Red Fox (*Vulpes vulpes*), in the Sierra Nevada, has heel pads with a bar across them; in general, its prints are larger and less clear (because of thick fur), its stride longer and its straddle narrower. A Domestic Cat (*Felis catus*) or Bobcat (p. 42) print will lack claw marks and have a larger, less symmetrical heel pad.

39

fore

hind

Fore Print
(hind print slightly smaller)
Length: 3.0–4.3 in (7.6–11 cm)
Width: 3.3–4.8 in (8.4–12 cm)
Straddle
8.0–12 in (20–30 cm)
Stride
Walking: 13–32 in (33–81 cm)
Bounding: to 12 ft (3.7 m)
Size
Height: 26–31 in (66–79 cm)
Length: 3.5–5.0 ft (1.1–1.5 m)
Weight
70–200 lb (32–91 kg)

walking (fast)

MOUNTAIN LION
(Puma, Cougar)
Felis concolor

   The Mountain Lion is shy, elusive and nocturnal in nature, so finding its tracks is usually the best that trackers can hope for. Spread widely but sparsely because of its need for a big home territory, this large cat is essential in keeping the deer population down. It avoids large open areas, such as those in the San Joaquin Valley and the desert regions.

   Mountain Lion prints tend to be wider than long and the retractable claws never register. In winter, thick fur makes the prints look much larger, and may obscure the two lobes on the front of the heel pad. When a Mountain Lion walks, the hind print will either directly register or double register on the larger fore print. As this cat's speed increases, the hind print will tend to fall ahead of the fore print. In snow, the thick, long tail may leave a dragline, which can obscure some of the print detail. When necessary, the Mountain Lion is capable of bounding quickly to catch prey.

Similar Species: A Bobcat's (p. 42) tracks may be confused with a juvenile Mountain Lion's, but a Bobcat sinks into the snow more, and its prints are usually clearer.

fore

hind

Fore Print
(hind print slightly smaller)
Length: 1.8–2.5 in (4.6–6.4 cm)
Width: 1.8–2.6 in (4.6–6.6 cm)

Straddle
4.0–7.0 in (10–18 cm)

Stride
Walking: 8.0–16 in (20–41 cm)
Running: 4.0–8.0 ft (1.2–2.4 m)

Size
(female is slightly smaller)
Height: 20–22 in (51–56 cm)
Length: 25–30 in (64–76 cm)

Weight
15–35 lb (6.8–16 kg)

walking

*trotting
to loping*

BOBCAT (Wildcat)
Lynx rufus

The Bobcat is widely distributed throughout California. As Bobcats are very adaptable animals, you will likely find their tracks anywhere from wild mountainsides to the chaparral, and even into residential areas. This stealthy hunter sits motionless in the secret of the night, waiting to pounce on its prey.

A Bobcat's heel pads have two lobes to the front and three to the rear. Fore prints especially show asymmetry. The hind print usually registers exactly on the larger fore print in the walking track. As a Bobcat picks up speed, its track becomes a trot pattern made of groups of two prints, the hind leading the fore. At even greater speeds, its track becomes a group of four prints in a lope pattern. A Bobcat's feet leave draglines in deep snow. Unlike the trails of wild dogs, the Bobcat's trail meanders. Half-buried scat along the trail is a sign of a Bobcat marking its territory.

Similar Species: A juvenile Mountain Lion (p. 40) has similar prints. Large Domestic Cat (*Felis catus*) prints may be confused with a juvenile Bobcat's prints, but Domestic Cats have a shorter stride and a narrower straddle and do not wander far from home, especially in winter (feral cats may also be found). Note: dog, coyote and fox prints show claw marks; the fronts of the footpads are singly lobed.

Fore and Hind Prints
Length: 1.0–1.4 in (2.5–3.6 cm)
Width: 1.0–1.4 in (2.5–3.6 cm)

Straddle
3.0–4.0 in (7.6–10 cm)

Stride
Walking: 3.0–6.0 in (7.6–15 cm)

Size (female slightly smaller)
Length: 24–32 in (61–81 cm)

Weight
1.5–2.5 lb (0.7–1.1 kg)

walking

RINGTAIL
(Cacomistle, Civet Cat, Miner's Cat)
Bassariscus astutus

This pretty cousin of the Raccoon (p. 46) is seldom seen. Found in the foothills of most southern Californian mountains, it avoids higher altitudes and low-elevation plains. The secretive Ringtail is strictly nocturnal and it rarely leaves any sign of its passage on the rocky terrain that it frequents. It usually travels under the cover of shrubs, adding to the difficulty of tracking this mammal. It never goes far from water.

A Ringtail's small, rounded prints show five toes; the partially retractable claws only occasionally register. Just behind the main pad of the fore print, a second pad may be evident. The common walking pattern is an alternating sequence of prints, where the hind registers on or close to the fore print. If you find a Ringtail's trail, it may lead you into rocky terrain, up a tree or to the animal's den.

Similar Species: Small weasel-family members (pp. 56–71) have similar prints, but a Ringtail has a different gait and habitat, and its fifth toe registers more often. Domestic Cat (*Felis catus*) prints never show five toes or a second pad.

fore

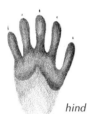

hind

Fore Print
Length: 2.0–3.0 in (5.1–7.6 cm)
Width: 1.8–2.5 in (4.6–6.4 cm)

Hind Print
Length: 2.4–3.8 in (6.1–9.7 cm)
Width: 2.0–2.5 in (5.1–6.4 cm)

Straddle
3.3–6.0 in (8.4–15 cm)

Stride
Walking/Running:
    7.0–20 in (18–51 cm)

Size (females slightly smaller)
Length: 24–37 in (61–94 cm)

Weight
11–35 lb (5.0–16 kg)

walking   *running group*

RACCOON
Procyon lotor

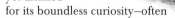

The inquisitive
Raccoon is
adored for
its distinc-
tive face
mask,
yet disliked
for its boundless curiosity—often
demonstrated with garbage cans. It is
common throughout most of the state, though it avoids the
driest regions of the Great Basin to the east. Look for its
tracks around waterbodies, in a diversity of habitats. Raccoons usually den up for the colder months.

The Raccoon's unusual print looks like a human
handprint, showing five well-formed toes. The small claws
appear as dots. Its highly dexterous forefeet rarely leave
heel prints, but its hind prints, which are generally much
clearer, do show heels. The Raccoon's peculiar walking
track shows a left fore print next to a right hind print and
vice versa. The fore print may be ahead of the hind print.
On the rare occasions when a Raccoon is out in deep snow,
it makes an alternating track. Raccoons occasionally run,
leaving clusters in which the two hind prints fall ahead of
the fore prints. Raccoons like to rest in trees.

Similar Species: The Fisher (p. 56) prefers deeper forests
and has different gaits. Unclear Opossum (p. 72) prints may
look similar, but the Opossum drags its tail.

47

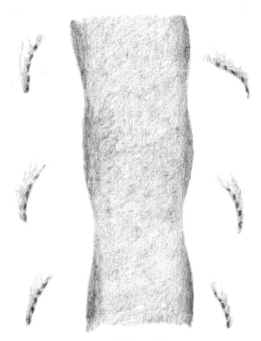

beach track

Size (male>female)
Length: 4.0–6.0 ft (1.2–1.8 m)
Weight
180–310 lb (82–140 kg)

HARBOR SEAL

Phoca vitulina

This common seal frequents the isolated beaches of the California coast. One of the smaller seals, it is a little shy, so it will usually slide off its rocky sentry post into the sea to escape curious naturalists. Look in sandy and muddy areas between rocky platforms for its tracks—their location and large size make them unmistakable. Sometimes a Harbor Seal will work its way up a river, so you may also find its tracks along riverbanks.

Seals are not the most graceful of animals on land, but they are unrivalled in the water. Their heavy, fat bodies and flipper-like feet can leave messy tracks: a wide trough (made by its cumbersome belly) with dents alongside (made as the seal pulled itself along with its forefeet). Look for the marks made by the nails.

Similar Species: The California Sea Lion (*Zalophus californianus*), which can weigh twice as much, also frequents the mainland coast. Steller's Sea Lion (*Eumetopias jubatus*) is sometimes seen along the shore too; it can weigh seven times as much as the little Harbor Seal. Other seals and sea lions can be found on the remote offshore islands.

fore

hind

Fore Print
Length: 2.5–3.5 in (6.4–8.9 cm)
Width: 2.0–3.0 in (5.1–7.6 cm)

Hind Print
Length: 3.0–4.0 in (7.6–10 cm)
Width: 2.3–3.3 in (5.8–8.4 cm)

Straddle
4.0–9.0 in (10–23 cm)

Stride
Walking/Running: 6.0–23 in (15–58 cm)

Size
(female is slightly smaller)
Length with tail: 3.0–4.3 ft (91–130 cm)

Weight
10–25 lb (4.5–11 kg)

running (fast)

RIVER OTTER
Lutra canadensis

No animal knows how to have more fun than a River Otter. If you are lucky enough to see otters at play, you will not soon forget the experience. Well-adapted for the aquatic environment, this otter is widespread along streams and other waterbodies through most of the Sierra Nevada and the San Joaquin Valley. Within an otter's home territory, you are sure to find a wealth of evidence along the riverbanks.

In soft mud, the River Otter's five-toed footprints show evidence of webbing, the hind prints more than the fore prints. The inner toes are set slightly apart. The heel may register, lengthening the print. Otter tracks are very variable: they show the typical two-print bounding of the weasel family, and, with faster runs, groups of four and three prints. The thick, heavy tail often leaves a mark over the prints. Otters love to slide in snow, often down riverbanks, leaving troughs nearly 12 inches (30 cm) wide. In summer they roll and slide on grass and mud.

Similar Species: Fisher (p. 56) prints are similar, but Fishers do not have webbed feet and do not leave conspicuous tail draglines and prefer forests; an otter in the forest is usually traveling directly between waterbodies.

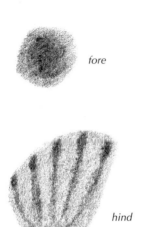

fore

hind

Fore Print
Length: 2.0–3.0 in (5.1–7.6 cm)
Width: 2.0–3.0 in (5.1–7.6 cm)

Hind Print
Length: up to 6.0 in (15 cm)
Width: up to 6.0 in (15 cm)

Straddle
to 12 in (30 cm)

Stride
Walking: to 15 in (38 cm)

Size (female is smaller)
Length with tail: 30–71 in (76–180 cm)

Weight
25–80 lb (11–36 kg)

walking

SEA OTTER
Enhydra lutris

As fun-loving as its freshwater cousin, the River Otter (p. 50), this large inhabitant of coastal waters is so suited to the sea that it seldom comes onto land. If you do catch sight of a Sea Otter on land, it might well be cavorting with sea lions, or waiting out a violent storm. If you are lucky, you may hear the tapping sound of an otter knocking on a shellfish with a stone to break it open. Sea Otter scat is recognizable by its crumbly texture and fragments of shell.

Although the Sea Otter rarely comes ashore, its tracks are most likely on beaches where kelp grows offshore. Like the seals, this otter is not as graceful on land as it is in the sea, because its webbed hind feet are almost flippers. The roundish fore print is much smaller than the hind print and the track will most likely be in an alternating walking pattern.

Similar Species: The Harbor Seal (p. 48) also has flipper-like feet, but it leaves an obvious trough with its belly and it has much larger forefeet.

53

fore

hind

Fore Print
Length with heel:
    4.0–7.5 in (10–19 cm)
Width: 4.0–5.0 in (10–13 cm)

Hind Print
Length: 3.5–4.0 in (8.9–10 cm)
Width: 4.0–5.0 in (10–13 cm)

Straddle
7.0–9.0 in (18–23 cm)

Stride
Walking: 3.0–12 in (7.6–30 cm)
Running: 10–40 in (25–100 cm)

Size (female is slightly smaller)
Height: 16 in (41 cm)
Length: 32–46 in (81–120 cm)

Weight
18–47 lb (8.2–21 kg)

running (slow)

WOLVERINE
Gulo gulo

The reputation of the robust and powerful Wolverine has earned it many nicknames, such as 'skunk bear' and 'Indian devil.' The Wolverine's need for pristine coniferous mountain forest wilderness has resulted in a scattered distribution at higher altitudes in the Sierra Nevada.

As with other members of the weasel family, the low, squat Wolverine has five toes, although the inner toe frequently does not register. While the fore print shows a small heel pad, the hind print rarely does. The Wolverine leaves a host of erratic trails, all typical of mustelids: an alternating walking pattern, the typical bounding print pairs and the common loping track of three- and four-print groups.

Similar Species: Other members of the weasel family (pp. 50–71) have smaller prints and less erratic tracks. When only four toes register, Wolverine prints may be mistaken for a large Domestic Dog's (*Canis familiaris*) prints, but the pad shapes are very different.

fore

Fore Print
Length: 2.1–3.9 in (5.3–9.9 cm)
Width: 2.1–3.3 in (5.3–8.4 cm)
Hind Print
Length: 2.1–3.0 in (5.3–7.6 cm)
Width: 2.0–3.0 in (5.1–7.6 cm)
Straddle
3.0–7.0 in (7.6–18 cm)
Stride
Walking: 7.0–14 in (18–36 cm)
Running: 1.0–4.2 ft (30–130 cm)
Size (male>female)
Length with tail:
    33–41 in (84–100 cm)
Weight
3.0–12 lb (1.4–5.4 kg)

walking   *bounding*

FISHER (Black Cat)
Martes pennanti

This agile hunter of the mixed-hardwood and coniferous forests of the Sierra Nevada is comfortable both on the ground and in the trees. Its speed and eager hunting antics make for exciting tracking as it races up trees and along the ground in its quest for squirrels. A Fisher's tracks may lead to the plucked remains of an unfortunate Porcupine (p. 82); the Fisher is one of the very few predators to regularly eat Porcupines.

The Fisher occasionally leaves a direct-registering alternating walking track. However, it usually prefers to bound in typical weasel fashion, leaving angled pairs of prints that represent the direct register of hind print over fore print. Though all five toes may register, the small inner toe frequently does not. Only the forefoot has a small heel pad that can show up in the print. Fisher tracks, which often vary within a short distance, are not associated with water–the name 'Fisher' is a misnomer!

Similar Species: Male Marten (p. 58) tracks may be confused with a small female Fisher's, but a Marten weighs less and leaves smaller, shallower prints. When only four toes register, Fisher prints may look like a Bobcat's (p. 42).

Fore and Hind Prints
Length: 1.8–2.5 in (4.6–6.4 cm)
Width: 1.5–2.8 in (3.8–7.1 cm)

Straddle
2.5–4.0 in (6.4–10 cm)

Stride
Walking: 4.0–9.0 in (10–23 cm)
Running: 9.0–46 in (23–120 cm)

Size (male>female)
Length with tail:
    21–27 in (53–69 cm)

Weight
1.5–2.8 lb (0.7–1.3 kg)

walking   *bounding*

MARTEN (Pine Marten)
Martes americana

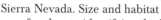

In southern California, this aggressive predator is found only in the montane and subalpine coniferous forests of the Sierra Nevada. Size and habitat are often key to identifying the Marten's tracks.

The Marten seldom leaves a clear print: the heel pad is very undeveloped and, in winter, the hairiness of the feet often obscures all pad detail, especially from the poorly developed palm pads. Often the inner toe fails to register, so the prints may show only four toes. In the Marten's alternating walking pattern, the hind print registers on the fore print. In a bounding track, the hind prints fall on the fore prints, in slightly angled groups of two, in the typical weasel-family pattern. Gallop groups may be three- or four-print clusters–see the River Otter (p. 50). If you follow the criss-crossing tracks, you may find that the Marten has scrambled up a tree; look for the sitzmark where it has jumped down again.

Similar Species: Female Fisher (p. 56) prints overlap in size with large male Marten prints, but Fishers leave clearer tracks. Male Mink (p. 60) prints overlap in size with small female Marten prints, but Mink do not climb trees and, unlike Martens, are often found near water.

fore left

hind left

Fore and Hind Prints
Length: 1.3–2.0 in (3.3–5.1 cm)
Width: 1.3–1.8 in (3.3–4.6 cm)
Straddle
2.1–3.5 in (5.3–8.9 cm)
Stride
Walking/Running: 8.0–35 in (20–89 cm)
Size
(female is slightly smaller)
Length with tail: 19–28 in (48–71 cm)
Weight
1.5–3.5 lb (0.7–1.6 kg)

bounding

MINK

Mustela vison

The lustrous Mink prefers watery habitats surrounded by brush or forest. In southern California, it is found only in parts of the Sierra Nevada. At home as much on land as in water, this nocturnal hunter can be an exciting animal to track. Like the River Otter (p. 50), the Mink likes to slide in snow, leaving a trough carved out by its body for an observant tracker to spot.

The fore print of the Mink shows five (sometimes four) toes with five loosely connected palm pads in an arc, but the hind print shows only four palm pads. The heel pad rarely shows on the fore print but may register on the hind print, lengthening it. The Mink prefers the typical weasel-family bounding pattern of double prints, consistently spaced and slightly angled. Nevertheless, its tracks show much diversity in gait and may also appear as an alternating walk, or as a run with three- and four-print groups, like the River Otter.

Similar Species: Small Martens (p. 58), with similar prints, do not have a consistent double-print bounding gait or live near water. Weasels (pp. 62–65) have similar tracks.

Fore and Hind Prints
Length: 1.1–1.8 in (2.8–4.6 cm)
Width: 0.8–1.0 in (2.0–2.5 cm)

Straddle
1.8–2.8 in (4.6–7.1 cm)

Stride
Bounding: 9.5–43 in (24–110 cm)

Size (male>female)
Length with tail:
    12–22 in (30–56 cm)

Weight
3.0–12 oz (85–340 g)

bounding

LONG-TAILED WEASEL
Mustela frenata

Weasels are energetic hunters, with an avid appetite for rodents. Following their tracks can reveal much about these nimble creatures' activities. Although weasels are active all year, their tracks are most evident in winter, when they frequently burrow into the snow or, in pursuit of their prey, enter an existing rodent hole. From time to time, weasel tracks may lead you up a tree; weasels have also been known to take to water. The largest weasel in southern California, the Long-tailed Weasel is widely distributed, except in very arid regions.

Because of a weasel's light weight and small, hairy feet, the pad detail is often obscured, especially in snow. Even with clear tracks, the inner (fifth) toe rarely registers. This weasel's tracks show typical bounding habits with an irregularity in stride, which is sometimes short and sometimes long, with no consistent behavior. In deep snow, look for holes where the weasel has suddenly plunged into the snow, perhaps in pursuit of prey.

Similar Species: A small female may leave tracks the same size as a large male Short-tailed Weasel's (p. 64). Mink (p. 60) tracks are similar.

Fore and Hind Prints
Length: 0.8–1.3 in (2.0–3.3 cm)
Width: 0.5–0.6 in (1.3–1.5 cm)

Straddle
1.0–2.1 in (2.5–5.3 cm)

Stride
Bounding: 9.0–35 in (23–89 cm)

Size (male>female)
Length with tail:
    8.0–14 in (20–36 cm)

Weight
1.0–6.0 oz (28–170 g)

bounding

SHORT-TAILED WEASEL
(Ermine)
Mustela erminea

    Smaller than the Long-tailed Weasel, this weasel reaches into southern California only in the Sierra Nevada. It prefers woodlands and meadows up to higher elevations, but does not favor wetlands or denser coniferous forests.

    Successful identification of which weasel made a particular track can be troublesome. Pay close attention to the straddle and stride, but note that small females of a larger species and large males of a smaller species may have similar tracks. Clues can be gained from the habit displayed in jumping patterns. This weasel's bounding track may show short strides alternating with long ones. Also check the distribution and the habitat.

Similar Species: A small female Long-tailed Weasel (p. 62) or a small Mink (p. 60) may leave tracks like a large male Short-tailed Weasel's.

fore

hind

**Fore Print
(hind print slightly shorter)**
Length: 2.5–3.0 in (6.4–7.6 cm)
Width: 2.3–2.8 in (5.8–7.1 cm)

Straddle
4.0–7.0 in (10–18 cm)

Stride
Walking: 6.0–12 in (15–30 cm)

Size
Length: 21–35 in (53–89 cm)

Weight
13–25 lb (5.9–11 kg)

walking

BADGER
Taxidea taxus

The squat shape and unmistakable face of this bold animal are most likely to be seen in open grasslands, though the Badger also ventures into higher mountain country; it is found throughout most of the state. Thick shoulders and forelegs, coupled with long claws, make for a powerful digging animal. Unlike most other weasel-family members, the Badger likes to den up in a hole during the really cold months of winter in more northerly regions, so look for its tracks in spring and fall snow.

All five toes on each foot register. A Badger's long claws are evident in the pigeon-toed track that it leaves as it waddles along, although the claws on the hind feet are not as long as those on the forefeet. When a Badger walks, the alternating track is a double register, with the hind print sometimes falling just behind the fore print and sometimes slightly ahead. The Badger's wide, low body will often plow through deeper snow and obscure the print detail.

Similar Species: In snow a Porcupine's (p. 82) tracks may be similar, but will show draglines made by its tail and quills and will likely lead up a tree, not to a hole.

fore

hind

Fore Print
Length: 1.5–2.2 in
    (3.8–5.6 cm)
Width: 1.0–1.5 in (2.5–3.8 cm)

Hind Print
Length: 1.5–2.5 in
    (3.8–6.4 cm)
Width: 1.0–1.5 in (2.6–3.8 cm)

Straddle
2.8–4.5 in (7.1–11 cm)

Stride
Walking/Running:
    2.5–8.0 in (6.4–20 cm)

Size
Length with tail:
    20–32 in (51–81 cm)

Weight
6.0–14 lb (2.7–6.4 kg)

walking (fast)

running

STRIPED SKUNK
Mephitis mephitis

This striking skunk has a notorious reputation for its vile smell; the lingering odor is often the best sign of its presence. The Striped Skunk lives in diverse habitats–but not deserts–and is widespread through the western and coastal counties of southern California.

Both fore and hind feet have five toes. The long claws on the forefeet often register. Smooth palm pads and small heel pads leave surprisingly small prints. Skunks mostly walk–with such a potent smell for their defense, and those memorable black and white stripes, they rarely need to run. Unlike other weasel-family members, skunks rarely show any consistent pattern in their tracks, but an alternating walking pattern may be evident. The greater the skunk's speed, the more its hind print oversteps the fore. Should a skunk need to run, its track is a closely set pattern of clumsy four-print groups. In snow, skunks drag their feet.

Similar Species: The Western Spotted Skunk (p. 70), with smaller prints in a very random pattern, is found in the same range, but also east of the Sierra Nevada and in the extreme southeast along the Colorado River.

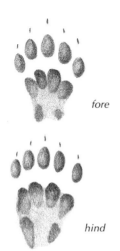

fore

hind

Fore Print
Length: 1.0–1.3 in (2.5–3.3 cm)
Width: 0.9–1.1 in (2.3–2.8 cm)
Hind Print
Length: 1.2–1.5 in (3.0–3.8 cm)
Width: 0.9–1.1 in (2.3–2.8 cm)
Straddle
2.0–3.0 in (5.1–7.6 cm)
Stride
Walking: 1.5–3.0 in (3.8–7.6 cm)
Jumping: 6.0–12 in (15–30 cm)
Size
Length: 13–25 in (33–64 cm)
Weight
0.6–2.2 lb (0.3–1.0 kg)

walking

running

WESTERN SPOTTED SKUNK (Civet Cat)
Spilogale gracilis

This beautifully marked skunk, smaller than its striped cousin, is found through most of southern California, except high in the Sierra Nevada. It enjoys diverse habitats, such as scrubland, forests and farmland, but it is a rare sight, because of its nocturnal habits and because it dens up in winter, coming out only on warmer nights.

This skunk leaves a very haphazard trail as it forages for food on the ground. Occasionally, with ease, it climbs trees. Long claws on the forefeet often register, and the palm and heel may leave defined pad marks. Although this skunk rarely runs, when it does so it may bound along, leaving groups of four prints, hind ahead of fore. It sprays only when truly provoked, so its powerful odor is less frequently detected than that of the Striped Skunk.

Similar Species: The Striped Skunk (p. 68), with a similar range, has larger prints and less-scattered tracks with a shorter running stride (or it jumps); it does not climb trees.

fore

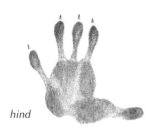

hind

Fore Print
Length: 2.0–2.3 in (5.1–5.8 cm)
Width: 2.0–2.3 in (5.1–5.8 cm)

Hind Print
Length: 2.5–3.0 in (6.4–7.6 cm)
Width: 2.0–3.0 in (5.1–7.6 cm)

Straddle
4.0–5.0 in (10–13 cm)

Stride
5.0–11 in (13–28 cm)

Size
Length: 2.0–2.5 ft (61–76 cm)

Weight
9.0–13 lb (4.1–5.9 kg)

walking

running

OPOSSUM
Didelphis virginiana

This slow-moving, nocturnal marsupial lives in coastal regions of the state. While the Opossum is found in many types of habitat, it shows a preference for open woodland or brushland around water-bodies. It is quite tolerant of farming and residential areas. Opossum tracks can often be seen in mud near the water; follow them and you might come across this strange animal playing dead ('playing 'possum') in the hope that you will leave it alone. If you find some roadkill, which the Opossum likes to feed on, look for tracks along the roadside.

Opossums are excellent climbers, so do not be surprised if their tracks lead to a tree. They have two walking habits: the common alternating pattern, with the hind prints registering on the fore prints, and a Raccoon-like paired-print pattern, with the hind print next to the opposing fore print. The long, very distinctive, inward-pointing thumb of the hind foot does not make a claw mark.

Similar Species: Prints in which the distinctive thumbs don't show, as in sand or fine snow, may be mistaken for a Raccoon's (p. 46).

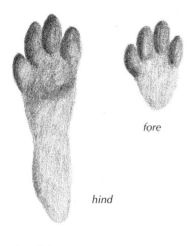

fore

hind

Fore Print
Length: 1.5–3.0 in (3.8–7.6 cm)
Width: 1.3–1.7 in (3.3–4.3 cm)

Hind Print
Length: 2.5–4.0 in (6.4–10 cm)
Length with heel: to 6.0 in (15 cm)
Width: 1.5–2.5 in (3.8–6.4 cm)

Straddle
4.0–7.0 in (10–18 cm)

Stride
Hopping: 5.0–10 ft (1.5–3.0 m)
In alarm: 20 ft (6.1 m)

Size
Length: 18–25 in (46–64 cm)

Weight
4.0–8.0 lb (1.8–3.6 kg)

hopping

BLACK-TAILED JACKRABBIT
Lepus californicus

This athletic hare frequents open and agricultural areas through much of the state, including desert country and sometimes at higher elevations. It can reach speeds of 35 mph (55 km/h). Often in groups, this nocturnal hare is infrequently seen.

Both fore and hind prints show four toes; the hind foot may often register a long heel when the hare walks slowly. When it hops, this hare creates print groups in a triangular pattern; as it speeds up, these print groups spread out considerably. Following the tracks could lead you to the hare's 'form'—a depression where it rests—or to an urgent zigzag pattern that indicates where the hare fled from danger. With its strong hind legs, it is capable of leaping up to 20 feet (6.1 m) to avoid predators.

Similar Species: The slightly larger White-tailed Jackrabbit (*Lepus townsendii*) of the Sierra Nevada may have similar tracks. The Snowshoe Hare (p. 76), found in more mountainous regions, spreads its hind toes more, takes shorter leaps and requires dense cover. Coyote (p. 34) prints resemble heel-less jackrabbit prints, but the gait is very different.

hind

fore

Fore Print
Length: 2.0–3.0 in (5.1–7.6 cm)
Width: 1.5–2.0 in (3.8–5.1 cm)
Hind Print
Length: 4.0–6.0 in (10–15 cm)
Width: 2.0–3.5 in (5.1–8.9 cm)
Straddle
6.0–8.0 in (15–20 cm)
Stride
Hopping: 0.8–4.2 ft (24–130 cm)
Size
Length: 12–21 in (30–53 cm)
Weight
2.0–4.0 lb (0.9–1.8 kg)

hopping

SNOWSHOE HARE (Varying Hare)

Lepus americanus

This hare is well known for its color change, from summer brown to winter white, and for its huge hind feet, which enable it to 'float' on the surface of snow. It is widespread in the coniferous forests of the Sierra Nevada and the White and Inyo mountains. It favors brushy areas along creeks, both for food and to hide from the Coyote (p. 34), its most likely predator. Well-worn runways are used as escape runs. Hares are most active at night.

As with rabbits and other hares, the Snowshoe Hare usually leaves a hopping track, with groups of four prints in a triangular pattern; they can be quite long if the hare is running quickly. A hare track's most distinctive feature is that the hind print is much larger than the fore print. In winter, heavy fur thickens the toes of a Snowshoe Hare's hind feet, which splay out to distribute its weight when it runs on snow. If you are lucky, you might even come across a resting hare, since they do not live in burrows. Signs of this hare's presence include twigs and stems that have been neatly severed at a 45° angle.

Similar Species: A cottontail (p. 78) is smaller. Though jackrabbits (p. 74), of the open country, are larger, the Snowshoe Hare has larger hind prints.

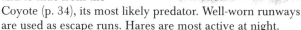

fore

hind

Fore Print
Length: 1.0–1.5 in (2.5–3.8 cm)
Width: 0.8–1.3 in (2.0–3.3 cm)
Hind Print
Length: 3.0–3.5 in (7.6–8.9 cm)
Width: 1.0–1.5 in (2.5–3.8 cm)
Straddle
4.0–5.0 in (10–13 cm)
Stride
Hopping: 7.0–36 in (18–91 cm)
Size
Length: 12–17 in (30–43 cm)
Weight
1.3–3.0 lb (0.6–1.4 kg)

hopping

DESERT COTTONTAIL
(Audubon's Cottontail)
Sylvilagus audubonii

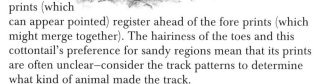

This abundant, common rabbit can be found anywhere from brushland to desert country. It often hides in the burrows of other animals for safety or rest.

As with other rabbits and hares, its most common track is a triangular grouping of four prints. The larger hind prints (which can appear pointed) register ahead of the fore prints (which might merge together). The hairiness of the toes and this cottontail's preference for sandy regions mean that its prints are often unclear—consider the track patterns to determine what kind of animal made the track.

Similar Species: The Mountain Cottontail (*Sylvilagus nuttallii*), on the eastern slopes of the Sierra Nevada, is similar in size. The Brush Rabbit (*Sylvilagus bachmanii*) of the coast is slightly smaller. Jackrabbits (p. 74) leave much larger print clusters and take longer strides. Squirrel (p. 92–99) tracks show a similar pattern, but with the fore prints more consistently side by side.

fore

hind

Fore Print
Length: 0.8 in (2.0 cm)
Width: 0.6 in (1.5 cm)

Hind Print
Length: 1.0–1.2 in (2.5–3.0 cm)
Width: 0.6–0.8 in (1.5–2.0 cm)

Straddle
2.5–3.5 in (6.4–8.9 cm)

Stride
Walking/Running: 4.0–10 in (10–25 cm)

Size
Length: 6.5–8.5 in (17–22 cm)

Weight
4.0–6.0 oz (110–170 g)

bounding

PIKA (Cony, Rock Rabbit)
Ochotona princeps

High up in the Cascade Range or the Sierra Nevada, you are more likely to hear the squeak of this cousin of the rabbits than to see it, as it is quick to disappear under the rocks for protection when alarmed. The Pika, confined to areas of high elevation, rarely leaves good tracks because it prefers exposed, rocky areas on mountain slopes–its tracks are most likely to be found in spring, on patches of snow or in mud. A more conspicuous sign of the Pika's presence are its little hay piles, set to dry in the sun for the long winter ahead, during which the Pika feeds on its stored food and remains active under the snow.

The Pika's fore print usually shows five toes, though one may not register. The hind print shows only four toes. Pika tracks may appear as an erratic alternating pattern or as three- and four-print bounding groups.

Similar Species: A rabbit or hare (pp. 74–79) may leave similar tracks, but has different habitats and never shows five toes on its fore prints. Moreover, the Pika, with smaller, rounder prints, does not have a long heel.

fore

hind

Fore Print
Length: 2.3–3.3 in (5.8–8.4 cm)
Width: 1.3–1.9 in (3.3–4.8 cm)

Hind Print
Length: 2.8–3.9 in (7.1–9.9 cm)
Width: 1.5–2.0 in (3.8–5.1 cm)

Straddle
5.5–9.0 in (14–23 cm)

Stride
Walking: 5.0–10 in (13–25 cm)

Size
Length with tail:
2.2–3.4 ft (67–100 cm)

Weight
10–28 lb (4.5–13 kg)

walking

PORCUPINE
Erethizon dorsatum

This notorious and easily recognized rodent rarely runs, because its many long quills are a formidable defense. Widespread through the coniferous forests of the Sierra Nevada and the San Bernadino Mountains, the Porcupine can also be seen in more open areas.

The most common Porcupine track is an alternating walking pattern, with the longer hind print registering on or slightly ahead of the fore print. Look for long claw marks on both prints. The fore print shows four toes and the hind print five. On clear prints, the unusual pebbly surface of the solid heel pads may show. However, a Porcupine's pigeon-toed footprints are often obscured by scratches from its heavy, spiny tail. A Porcupine's waddling gait shows in its track. In deeper snow, this squat animal drags its feet, and it may leave a trough with its body. A Porcupine's trail might lead you to a tree, where these animals spend much of their time feeding—look for chewed bark or nipped buds lying on the forest floor.

Similar Species: The Badger (p. 66) also has pigeon-toed prints, but it does not drag its tail or climb trees.

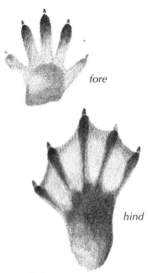

fore

hind

Fore Print
Length: 2.5–4.0 in (6.4–10 cm)
Width: 2.0–3.5 in (5.1–8.9 cm)

Hind Print
Length: 5.0–7.0 in (13–18 cm)
Width: 3.3–5.3 in (8.4–13 cm)

Straddle
6.0–11 in (15–28 cm)

Stride
Walking: 3.0–6.5 in (7.6–17 cm)

Size
Length with tail: 3.0–3.9 ft (91–120 cm)

Weight
28–75 lb (13–34 kg)

walking

BEAVER
Castor canadensis

Few animals leave as many signs of their presence as the Beaver, the largest North American rodent and a common sight around water. It has a scattered distribution, most notably in the Coast Ranges and eastwards to the mountains east of Los Angeles, and along the Colorado River. Signs of Beaver activity include dams and domed lodges made of sticks and branches, and the stumps of felled trees–trunks gnawed clean of bark bear marks of the Beaver's huge incisors. Beavers also make scent mounds marked with castoreum, a strong-smelling yellowish fluid.

A Beaver's tracks are often made less clear by its thick, scaly tail, or by the branches that it drags about for construction and food. If you find clear tracks, check the large hind prints for signs of webbing and broad toenails–the inner second toenail usually does not show. Rarely do all five toes on each foot register. The Beaver's track may be in an irregular alternating sequence, often with a double register. Beavers frequently make well-worn trails.

Similar Species: Little confusion arises, because the Beaver leaves so many distinctive signs of its presence.

fore

hind

walking

Fore Print
Length: 1.1–1.5 in (2.8–3.8 cm)
Width: 1.1–1.5 in (2.8–3.8 cm)
Hind Print
Length: 1.6–3.2 in (4.1–8.1 cm)
Width: 1.5–2.1 in (3.8–5.3 cm)
Straddle
3.0–5.0 in (7.6–13 cm)
Stride
Walking: 3.0–5.0 in (7.6–13 cm)
Running: to 1.0 ft (30 cm)
Size
Length with tail: 16–25 in (41–64 cm)
Weight
2.0–4.0 lb (0.9–1.8 kg)

MUSKRAT
Ondatra zibethica

This rodent is found where there is water, in the Central Valley and at higher elevations of the Sierra Nevada, as well as along the Colorado River. Beavers (p. 84) are very tolerant of Muskrats and even allow them to live in parts of their lodges. Muskrats are active all year, and they leave plenty of signs of their presence. They dig an extensive network of burrows, often undermining the riverbank, so do not be surprised if you suddenly fall into a hidden hole! Other signs of this rodent are their small lodges in the water, and the beds of vegetation on which they rest, sun and feed during the summer.

The reduced inner toe of the five on each forefoot rarely registers in the print, but the hind print shows five well-formed toes. The prints are usually in an alternating track, with the hind print just behind or slightly overlapping the fore print. In snow, a Muskrat's feet drag a lot, and its tail leaves a sweeping dragline.

Similar Species: Few animals share this water-loving rodent's habits.

fore

hind

walking

Fore and Hind Prints
Length: 1.6–2.0 in (4.1–5.1 cm)
Width: 1.0 in (2.5 cm)
Straddle
2.0–3.0 in (5.1–7.6 cm)
Stride
Walking: 3.0 in (7.6 cm)
Size
Length: 9.5–19 in (24–48 cm)
Weight
1.0–3.0 lb (0.5–1.4 kg)

MOUNTAIN BEAVER
(Aplodontia)
Aplodontia rufa

This stocky rodent with the inconspicuous tail is a curious resident of the forests of the Sierra Nevada, but only in the north of southern California. The slow-moving Mountain Beaver is neither truly a beaver nor a montane animal—it is an enthusiastic burrower in moist soils. Other signs of its activity are more noticeable than its tracks: for instance, its large piles of hay on logs or on the ground and its nearby labyrinth of shallow tunnels. After the snow melts in spring, look for soil cores much like a pocket gopher's (p. 106), only larger. Trees and shrubs may have their shoot ends neatly nipped off; sometimes its appetite for clipping leaders or peeling bark off conifers can cause trees serious damage.

Both fore and hind prints are narrow and show five toes, though the inner toe of the forefoot only partially registers. The hind print shows a slightly longer heel. In the Mountain Beaver's alternating walking track, the hind print occasionally registers on the fore print.

Similar Species: There are few animals of similar size with similar activities and prints in dark, moist forests.

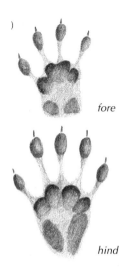

fore

hind

Fore and Hind Prints
Length: 1.5–2.5 in (3.8–6.4 cm)
Width: 1.0–1.5 in (2.5–3.8 cm)

Straddle
3.0–5.0 in (7.6–13 cm)

Stride
Walking: 2.0–6.0 in (5.1–15 cm)
Running: 3.0–14 in (15–36 cm)

Size (male>female)
Length with tail:
    17–28 in (43–71 cm)

Weight
5.0–10 lb (2.3–4.5 kg

walking *running*

YELLOW-BELLIED MARMOT
Marmota flaviventris

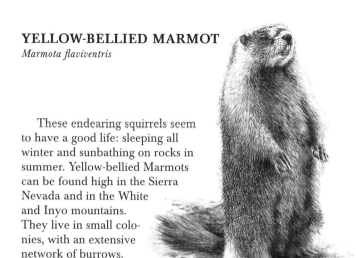

These endearing squirrels seem to have a good life: sleeping all winter and sunbathing on rocks in summer. Yellow-bellied Marmots can be found high in the Sierra Nevada and in the White and Inyo mountains. They live in small colonies, with an extensive network of burrows. Marmots are a joy to watch when they play-fight.

The fore print has four toes with three palm pads and two heel pads that are not always evident. The hind print has five toes, four palm pads and two poorly registering heel pads. In a marmot's usual alternating walking pattern, its hind print registers over its fore print. When a marmot runs, its track shows a group of four prints, its two hind prints ahead of its fore prints. Because of this marmot's preference for rocky habitats, tracks may be hard to come by, but they can be found in spring and fall snowfalls.

Similar Species: A small Raccoon's (p. 46) running track may be confused with a Yellow-bellied Marmot's, but a Raccoon shows five toes on each fore print; consider habitat and behavior to reduce confusion.

fore

hind

Fore Print
Length: 1.0–1.8 in (2.5–4.6 cm)
Width: 1.0 in (2.5 cm)

Hind Print
Length: 2.3–3.0 in (5.8–7.6 cm)
Width: 1.1–1.5 in (2.8–3.8 cm)

Straddle
3.8–6.0 in (9.7–15 cm)

Stride
Running: 8.0–36 in (20–91 cm)

Size
Length with tail: 20–23 in (51–58 cm)

Weight
12–32 oz (340–910 g)

running

WESTERN GRAY SQUIRREL
(California Gray Squirrel)
Sciurus griseus

This large, familiar squirrel lives along the Coast Ranges and through the Sierra Nevada. It can be a common sight in woodlands and in orchards, where its frequent raids on nut trees make it unpopular. In winter it can leave a wealth of evidence as it scurries about digging up nuts that it buried in the previous fall.

The Western Gray Squirrel leaves a typical squirrel track when it runs or bounds. The hind prints fall slightly ahead of the fore prints. A clear fore print shows four toes with sharp claws, four fused palm pads and two heel pads. The hind print shows five toes and four palm pads; if the full length of the heel registers, it shows two small heel pads.

Similar Species: The introduced Eastern Gray Squirrel (*S. carolinensis*), slightly smaller, with a reddish brown belly, is common in parks and residential areas. The russet Fox Squirrel (*S. niger*) is also common, in parks and orchards. A rabbit's or hare's (pp. 74–79) prints make a longer pattern, and its fore prints rarely register side by side when it runs. Chipmunk (p. 100), Douglas Squirrel (p. 94) and Northern Flying Squirrel (p. 96) tracks show smaller straddle and smaller prints in a similar pattern.

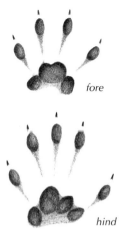

fore

hind

Fore Print
Length: 0.8–1.5 in (2.0–3.8 cm)
Width: 0.5–1.0 in (1.3–2.5 cm)

Hind Print
Length: 1.5–2.3 in (3.8–5.8 cm)
Width: 0.8–1.3 in (2.0–3.3 cm)

Straddle
3.0–4.5 in (7.6–11 cm)

Stride
Running: 8.0–30 in (20–76 cm)

Size
Length with tail:
    11–14 in (28–36 cm)

Weight
5.3–11 oz (150–310 g)

bounding

*bounding
(in deep snow)*

DOUGLAS SQUIRREL
(Pine Squirrel, Chickaree)
Tamiasciurus douglasii

When you enter its territory in the coniferous forests of the Sierra Nevada, a Douglas Squirrel greets you with a loud, chattering call. Another obvious sign of this widespread squirrel is the large middens–piles of cone scales and cores–at the bases of trees that indicate favorite feeding sites.

Active year-round in their small territories, Douglas Squirrels make an abundance of tracks that lead from tree to tree or down a burrow. These energetic animals mostly run, with a gait that leaves groups of four prints; the five-toed hind prints fall ahead of the four-toed fore prints, which are often side by side, and the heels often do not register. In deeper snow, a Douglas Squirrel's prints merge to form diamond-shaped pairs.

Similar Species: The larger Western Gray Squirrel (p. 92) shares the same range. A Desert or Mountain cottontail's (p. 78) fore prints rarely register side by side when it runs. Chipmunk (p. 100) tracks and Northern Flying Squirrel (p. 96) tracks have smaller straddles and smaller prints.

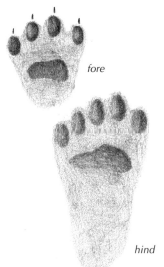

fore

hind

sitzmark into bounding

Fore Print
Length: 0.5–0.8 in (1.3–2.0 cm)
Width: 0.5 in (1.3 cm)

Hind Print
Length: 1.3–1.8 in (3.3–4.6 cm)
Width: 0.8 in (2.0 cm)

Straddle
3.0–3.8 in (7.6–9.7 cm)

Stride
Running: 11–29 in (28–74 cm)

Size
Length with tail: 9.0–12 in (23–30 cm)

Weight
4.0–6.5 oz (110–180 g)

NORTHERN FLYING SQUIRREL
Glaucomys sabrinus

This soft-furred brown acrobat is found in coniferous forests at quite high altitudes in the Sierra Nevada, with a small population in the San Bernadino Mountains. Its ideal habitat is widely spaced forest, especially with fir trees, where it can glide from tree to tree through the night, using the membranous flap between its legs. During the colder months of winter, Northern Flying Squirrels will den up together in a tree cavity for warmth.

Because of its gliding ability, this squirrel does not leave as many tracks as the Douglas Squirrel and it is very difficult to find any evidence of its presence in summer. In winter, however, you might come across a sitzmark–the distinctive shape that a squirrel makes when it lands on the ground–and a short bounding track, made as the squirrel rushed off to the nearest tree or to do some quick foraging. The bounding tracks left in snow are typical of squirrels and other rodents, with the hind prints falling ahead of the fore prints, which usually register side by side.

Similar Species: Douglas Squirrels (p. 94) usually have larger prints and never leave sitzmarks, but in deep snow their tracks resemble a Northern Flying Squirrel's. Chipmunks (p. 100) have a smaller straddle and smaller prints.

fore

hind

Fore Print
Length: 1.0–1.3 in (2.5–3.3 cm)
Width: 0.5–1.0 in (1.3–2.5 cm)

Hind Print
Length: 1.5–2.5 in (3.8–6.4 cm)
Width: 0.8–1.5 in (2.0–3.8 cm)

Straddle
2.0–4.0 in (5.0–10 cm)

Stride
Running: 7.0–18 in (18–46 cm)

Size
Length with tail: 14–20 in (36–51 cm)

Weight
10–26 oz (280–740 g)

bounding

CALIFORNIA GROUND SQUIRREL
Spermophilus beecheyi

The widespread California Ground Squirrel enjoys open terrain, especially short-cropped pastures and roadsides. Also found in rocky areas and open hillsides at higher elevations, it avoids the arid parts of the Great Basin. Though younger ground squirrels may remain active year-round, most hibernate during the winter, so tracks tend to be more evident around the many burrow entrances after late or early snowfalls and when it's muddy.

The small fifth toe of the forefoot rarely registers, but the two heel pads sometimes show. The larger hind foot registers five toes. Both fore and hind prints often show claw marks. Ground squirrels usually leave a typical squirrel track, with the hind prints registering ahead of the fore prints, which are usually placed diagonally.

Similar Species: The White-tailed Antelope Squirrel (*Ammospermophilus leucurus*) and the Round-tailed Ground Squirrel (*S. tereticaudus*) are found in the arid regions of the east. Nelson's Antelope Squirrel (*A. nelsoni*) likes grassy areas in the southern San Joaquin Valley. Chipmunk (p. 100) tracks are smaller. A tree squirrel's (pp. 92–97) tracks have a more square-shaped bounding group.

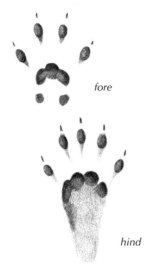

fore

hind

Fore Print
Length: 0.8–1.0 in (2.0–2.5 cm)
Width: 0.4–0.8 in (1.0–2.0 cm)

Hind Print
Length: 0.7–1.3 in (1.8–3.3 cm)
Width: 0.5–0.9 in (1.3–2.3 cm)

Straddle
2.0–3.1 in (5.1–7.9 cm)

Stride
Running: 7.0–15 in (18–38 cm)

Size
Length with tail: 9.0–10 in (23–25 cm)

Weight
2.0–3.7 oz (57–100 g)

bounding

MERRIAM'S CHIPMUNK

Tamias merriami

Of northern California's many delightful chipmunks, one of the more widespread is Merriam's Chipmunk, a dull-colored inhabitant in and above the chaparral and forests of the lower mountains. It can be found in the Coast Ranges at lower elevations, up through the western slopes of the Sierra Nevada and down through the Peninsular Ranges. In this warm climate, it is active for most of the year.

Chipmunks are so light that their prints rarely show fine details. They bound on their toes, so their forefoot heel pads often don't register and the hind feet have no heel pads. Forefeet have four toes; hind feet have five. The hind prints register ahead of the fore prints. A chipmunk's erratic trail often leads to extensive burrows.

Similar Species: The Lodgepole (*T. speciosus*) Chipmunk also lives in the Sierra Nevada and in the San Bernadino Mountains. The Panamint (*T. panamintinus*), Uinta (*T. umbrinus*) and Least (*T. minimus*) chipmunks occur in Inyo and Mono counties. The California Chipmunk (*T. obscurus*) is found in the Peninsular Ranges. Squirrels (pp. 92–99) unsually have larger prints. Mice (pp. 110–115) prints are smaller.

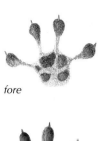

fore

hind

Fore Print
Length: 0.6–0.8 in (1.5–2.0 cm)
Width: 0.4–0.5 in (1.0–1.3 cm)

Hind Print
Length: 1.0–1.5 in (2.5–3.8 cm)
Width: 0.6–0.8 in (1.5–2.0 cm)

Straddle
2.3–2.7 in (5.8–6.9 cm)

Stride
Walking: 1.8–3.0 in (4.6–7.6 cm)
Jumping: 5.0–8.0 in (13–20 cm)

Size
Length with tail:
    9.0–15 in (23–38 cm)

Weight
3.9–6.0 oz (110–170 g)

walking   *running*

DESERT WOODRAT
Neotoma lepida

This nocturnal woodrat thrives in the arid regions of the Great Basin, up through the Coast Ranges and at lower elevations in the southern Sierra Nevada. The trail of a Desert Woodrat may lead you to a distinctive mass of cactus spines and sharp twigs overtop a burrow in which it nests and stores food. This woodrat is fond of spiny meals too, and it often strews the remains of the cacti that it relishes around the entrance to its house.

Four toes register on the fore print and five on the hind. The short claws rarely show. A woodrat often walks in an alternating fashion, with the hind print directly registering on the fore print. This woodrat also frequently runs, leaving a pattern of four prints with the larger hind prints registering ahead of the diagonally placed fore prints. This animal's stride tends to be short relative to the size of its feet.

Similar Species: The larger White-throated Woodrat (*N. albigula*) lives in the extreme southeast. The Dusky-footed Woodrat (*N. fuscipes*) prefers the forests and scrubland of the west and eastwards up into the mountains. A Norway Rat (p. 104) is usually found close to human activity. A Yellow-bellied Marmot (p. 90) has similar, much larger prints.

fore

hind

Fore Print
Length: 0.7–0.8 in (1.8–2.0 cm)
Width: 0.5 in (1.3 cm)

Hind Print
Length: 1.0–1.3 in (2.5–3.3 cm)
Width: 0.8–1.0 in (2.0–2.5 cm)

Straddle
3.0 in (7.6 cm)

Stride
Walking: 1.5–3.5 in (3.8–8.9 cm)
Jumping: 5.0–12 in (13–30 cm)

Size
Length with tail:
    13–19 in (33–48 cm)

Weight
7.0–18 oz (200–510 g)

walking

NORWAY RAT
(Brown Rat)
Rattus norvegicus

This despised rat is widespread almost anywhere humans have decided to build homes. Although it is not entirely dependent on people, it is rarely found in the wild. It is active both day and night.

The Norway Rat commonly leaves an alternating walking pattern, with the larger hind print registering close to or on the fore; the hind heel does not show. The fore print shows four toes, while the hind print shows five. When it runs, this colonial rat leaves groups of four prints with the diagonally placed fore prints registering behind the hind prints. In snow, the rat's tail often leaves a drag-line. Since rats live in groups, you may find that there are many tracks close together, often leading to their 2-inch (5.1 cm) wide burrows in the ground.

Similar Species: Woodrat (p. 102) tracks may be similar, but woodrats rarely associate with human activity, except in abandoned buildings. The Roof Rat or Black Rat (*Rattus rattus*), as widespread as the Norway Rat, is smaller.

fore

hind

Fore Print
Length: 1.0 in (2.5 cm)
Width: 0.6 in (1.5 cm)
Hind Print
Length: 1.0–1.5 in (2.5–3.8 cm)
Width: 0.5 in (1.3 cm)
Straddle
1.5–2.0 in (3.8–5.1 cm)
Stride
Walking: 1.3–2.0 in (3.3–5.1 cm)
Size (male>female)
Length with tail: 6.5–11 in (17–28 cm)
Weight
2.5–8.0 oz (71–230 g)

walking

BOTTA'S POCKET GOPHER
(Valley Pocket Gopher)
Thomomys bottae

This seldom-seen rodent can be found throughout southern California, though not at higher elevations in the Sierra Nevada. It spends most of its life in burrows, but from time to time it will venture out to move soil about or to find a mate. Pocket gophers prefer the soft, moist soils of mountain meadows but inhabit even very dry areas. The best sign of Botta's Pocket Gopher is the muddy mounds and tunnel cores, especially evident after spring thaw. Each mound marks the entrance to a burrow, which is always blocked with a plug. Look nearby to find tracks.

Both fore and hind feet have five toes. The forefeet have well-developed long claws for digging, though it is a rare track that shows this much detail. The typical track is an alternating walk, where the hind print registers on or slightly behind the fore print.

Similar Species: The Mountain Pocket Gopher (*T. monticola*) prefers higher elevations in the Sierra Nevada. The association of pocket gopher tracks with their distinctive burrows leaves little room for confusion with other animals.

fore

hind

Fore Print
Length: 0.5 in (1.3 cm)
Width: 0.5 in (1.3 cm)
Hind Print
Length: 0.6 in (1.5 cm)
Width: 0.5–0.8 in (1.3–2.0 cm)
Straddle
1.3–2.0 in (3.3–5.1 cm)
Stride
Walking: 0.8 in (2.0 cm)
Running/Hopping:
    2.0–6.0 in (5.1–15 cm)
Size
Length with tail:
    6.3–8.4 in (16–21 cm)
Weight
1.5–3.0 oz (43–85 g)

walking

*hopping
(in snow)*

CALIFORNIA VOLE
Microtus californicus

There are so many vole species that positive track identification is hard, though considering the habitat can help. The California Vole, southern California's most widespread vole, is commonly found in damp meadows and alfalfa plantings in the southern coastal counties and on the western slopes of the Sierra Nevada. Small populations can also be found in the White and Inyo mountains.

Vole fore prints have four toes and hind prints five, but the prints are seldom clear. In a vole's paired alternating walking track, the hind print may register on the fore; it always does so in a vole's preferred hopping gait. Voles stay under the snow in winter, so look for distinctive piles of cut grass from their ground nests and for tiny teeth marks in the bark at shrub bases. In summer, vole paths become little runways through the grass.

Similar Species: The Long-tailed Vole (*M. longicaudus*) can be found in the Sierra Nevada and the San Bernadino Mountains. The Montane Vole (*M. montanus*) is restricted to the Sierra Nevada. The Sagebush Vole (*Lagurus curtatus*) inhabits the arid Great Basin in Inyo and Mono counties. A Deer Mouse's (p. 112) paired hop pattern has a shorter stride.

hind

fore

slow hopping group (in sand)

Fore Print
Length: 0.3 in (0.8 cm)
Width: 0.3 in (0.8 cm)

Hind Print
Length: 0.4 in (1.0 cm)
    with heel 1.1 in (2.8 cm)
Width: 0.4 in (1.0 cm)

Straddle
1.5 in (3.8 cm)

Stride
Running: 0.8–4.5 in (2.0–11 cm)

Size
Length with tail: 7.4–9.2 in (19–23 cm)

Weight
0.6–0.8 oz (17–23 g)

bounding

CALIFORNIA POCKET MOUSE
Chaetodipus californicus

A number of endearing pocket mice can be found scattered through southern California. The nocturnal California Pocket Mouse is a common resident of the coastal counties and western slopes of the Sierra Nevada, where it inhabits grassy areas with fine soil and forages near the cover of low shrubs. In the extremes of winter or summer, this rodent can enter a state of torpor to sleep out the harsher times.

As a pocket mouse bounds along, it leaves four-print groups: the two fore prints fall closely side by side and the larger, wider-set hind feet overstep. The trail is seldom clear. Look for the little depressions in which pocket mice take dust baths to get rid of ticks and mites.

Similar Species: The Long-tailed Pocket Mouse (*Perognathus formosus*) is abundant in the desert counties. The Little Pocket Mouse (*P. longimembris*) has a scattered distribution at low elevations. The San Joaquin Pocket Mouse (*P. inornatus*) occurs in the Central Valley. The San Diego Pocket Mouse (*P. fallax*) is found in the southernmost counties. Deer Mouse (p. 112) tracks are similar but larger.

running group

Fore Print
Length: 0.3 in (0.8 cm)
Width: 0.3 in (0.8 cm)
Hind Print
Length: 0.6 in (1.5 cm)
Width: 0.4 in (1.0 cm)
Straddle
1.4–1.8 in (3.6–4.6 cm)
Stride
Running: 2.0–5.0 in
    (5.1–13 cm)
Size
Length with tail:
    6.0–9.0 in (15–23 cm)
Weight
0.5–1.3 oz (14–37 g)

running

*running
(in snow)*

DEER MOUSE

Peromyscus maniculatus

One of many mice in the region, this highly adaptable rodent might be encountered almost anywhere, except at the highest elevations. It may enter buildings, where it remains active during winter.

In perfect, soft mud, Deer Mouse fore prints each show four toes, three palm pads and two heel pads, and the hind prints show five toes and three palm pads; the hind heel pads rarely register. Running tracks—most noticeable in snow—show the hind prints in front of the close-set fore prints. The tracks may lead you up a tree or down into a burrow. In soft snow, the prints may merge and appear as larger pairs of prints, with tail drag evident.

Similar Species: The state-wide Western Harvest Mouse (*Reithrodontomys megalotis*) prefers grassy areas. The larger Brush Mouse (*P. boylii*) prefers the brushland of lower mountains and avoids arid regions. The House Mouse (*Mus musculus*) is more associated with humans. The California (Parasitic) Mouse (*P. californicus*) inhabits the coastal counties and the western slopes of the Sierra Nevada. A vole's (p. 108) merged two-print pattern has longer strides. A Western Jumping Mouse (p. 114) may have similar tracks. Chipmunks (p. 100) have a wider straddle. Shrews (p. 118) and pocket mice (p. 110) have a narrower straddle.

jumping group

Fore Print
Length: 0.3–0.5 in (0.8–1.3 cm)
Width: 0.3–0.5 in (0.8–1.3 cm)

Hind Print
Length: 0.5–1.3 in (1.3–3.3 cm)
Width: 0.5 in (1.3 cm)

Straddle
1.8–1.9 in (4.6–4.8 cm)

Stride
Hopping: 2.0–7.0 in (5.1–18 cm)
In alarm: 3.0–4.0 ft (91–120 cm)

Size
Length with tail: 7.0–9.0 in (18–23 cm)

Weight
0.6–1.3 oz (17–37 g)

jumping

WESTERN JUMPING MOUSE

Zapus princeps

Congratulations if you find and successfully identify the tracks of a Western Jumping Mouse! This hard-to-find rodent can be found in the Sierra Nevada, often in tall-grass meadows. The Western Jumping Mouse's preference for lush meadows and dense undergrowth and its long, deep hibernation period (about six months!) makes locating tracks very difficult.

Jumping mouse tracks are distinctive if you do find them. The two smaller fore prints register between the long hind feet. The long heels do not always show. When they jump, these mice make short leaps. The tail may leave a dragline in soft mud or unseasonable snow. An abundant sign of this rodent is the clusters of cut grass stems, about 5 inches (13 cm) long, lying in the meadows.

Similar Species: A Deer Mouse (p. 112) track may have the same straddle. A kangaroo rat (p. 116) also jumps, but only on its hind feet, not all four.

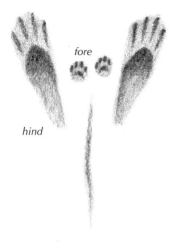

fore

hind

hopping group

Hind Print (fore print is much smaller)
Length: 1.3–1.6 in (3.3–4.1 cm)
Width: 0.5–0.8 in (1.3–2.0 cm)
Straddle
1.3–2.3 in (3.3–5.8 cm)
Stride
Hopping: 5.0–24 in (13–61 cm)
Leaping: 2.0–6.0 ft (61–180 cm)
Size
Length with tail: 8.6–10 in (22–25 cm)
Weight
1.3–1.8 oz (37–51 g)

hopping

MERRIAM'S KANGAROO RAT

Dipodomys merriami

This athletic rodent likes dry soils suitable for burrowing. This kangaroo rat, one of the smallest, is found east and south of the Sierra Nevada, but not near the coast. In cold weather it stays in its burrow, venturing out on milder nights. Though good tracks are hard to find, you may see some in sand; habit best identifies the track. Also look for the little depressions where these animals sand-bathe.

When a kangaroo rat hops slowly, its two small forefeet register between its large hind feet, which show long heel marks, and its long tail leaves a dragline. At speed, the fore prints do not register, the hind heel appears shorter and the tail only sometimes registers. Kangaroo rats can have either four or five toes per hind foot. Tap your fingers beside a burrow and you may hear thumping in reply.

Similar Species: The Desert Kangaroo Rat (*D. deserti*) shares the same range. The Pacific Kangaroo Rat (*D. agilis*) is found in the coastal counties from Point Conception southward. The Panamint Kangaroo Rat (*D. panamintinus*) inhabits the eastern slopes of the Sierra Nevada. The Chisel-toothed Kangaroo Rat (*D. microps*) is confined to the eastern desert counties. Heerman's Kangaroo Rat (*D. heermanni*) can be found in the Coast Ranges and the Sierra Nevada. The Western Jumping Mouse (p. 114) prefers lush areas.

running group

running

Fore Print
Length: 0.2 in (0.5 cm)
Width: 0.2 in (0.5 cm)

Hind Print
Length: to 0.5 in (1.3 cm)
Width: 0.3 in (0.8 cm)

Straddle
0.8–1.3 in (2.0–3.3 cm)

Stride
Running: 1.2–2.0 in (3.0–5.1 cm)

Size
Length with tail:
    3.3–4.3 in (8.4–11 cm)

Weight
0.1–0.3 oz (3–8 g)

ORNATE SHREW
Sorex ornatus

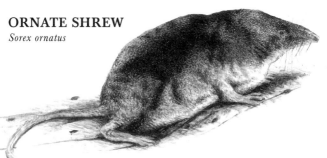

There are many tiny, frenetic shrews in southern California. If you find shrew tracks, the widespread Ornate Shrew is a likely candidate, though it is absent from the high Sierra Nevada and the arid eastern counties. It is difficult to observe closely, because of its rapid activity and its preference for thick undergrowth, often near wet areas.

In its energetic and unending quest for food, a shrew usually leaves a running pattern of four prints, but may slow to an alternating walking pattern. The individual prints in a group are often indistinct, but in mud or shallow, wet snow, you can even count the five toes on each print. In snow, a shrew's tail often leaves a dragline. If the shrew tunnels under the snow, it may leave a ridge of snow on the surface. A shrew's trail may disappear down a burrow.

Similar Species: Trowbridge's Shrew (*S. trowbridgii*) inhabits the Coast Ranges from Santa Barbara northwards. The tiny Desert Shrew (*Notiosorex crawfordi*) lives in the south. The Montane Shrew (*S. monticolus*) is found high in the Sierra Nevada and in the San Bernadino Mountains. The larger Water Shrew (*S. palustris*) lives in the Sierra Nevada. Mice (pp. 110–115) fore prints show four toes.

A molehill of the Broad-footed Mole

Some molehills and ridges of the Broad-footed Mole

Size
Length with tail:
5.2–7.5 in (13.2–19 cm)
Weight
1–2.5 oz (28–71g)

BROAD-FOOTED MOLE

Scapanus latimanus

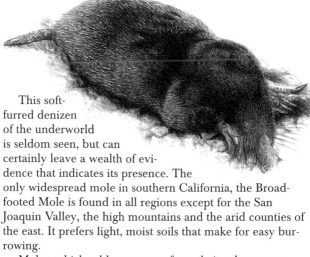

This soft-
furred denizen
of the underworld
is seldom seen, but can
certainly leave a wealth of evidence that indicates its presence. The
only widespread mole in southern California, the Broad-
footed Mole is found in all regions except for the San
Joaquin Valley, the high mountains and the arid counties of
the east. It prefers light, moist soils that make for easy burrowing.

Moles, which seldom emerge from their subterranean
existence, create an extensive network of burrows through
which they forage. Most of us are familiar with the hills that
form where the mole gets rid of excess soil from its burrows
and for which moles are frequently considered to be pests
when they mess up fine lawns. Occasionally, ridges on the
surface indicate where the burrows are; they will lead you
to the molehills scattered about.

Similar Species: The Shrew-mole (*Neurotrichus gibsii*),
confined to the moist coastal region of the Santa Cruz
Mountains, is the smallest North American mole and the
only other mole in southern California.

121

BIRDS, AMPHIBIANS AND REPTILES

A guide to the animal tracks of southern California is not complete without some consideration of the birds, amphibians and reptiles in the region. Only a few have been chosen as examples.

The differences among bird species are not necessarily reflected in their tracks, so several bird species have been chosen to represent the main types common to this region.

Bird tracks can often be found in abundance in snow and are clearest in shallow, wet snow. The shores of streams and lakes are very reliable locations to find bird tracks—the mud there can hold a clear print for a long time. The sheer number of tracks made by shorebirds and waterfowl can be astonishing. While some bird species prefer to perch in trees or soar across the sky, it can be entertaining to follow the tracks of those that do spend a lot of time on the ground. They can spin around in circles and lead you in all directions. The track may suddenly end as the bird takes flight, or it might terminate in a pile of feathers, the bird having fallen victim to a hungry predator.

Many amphibians and turtles depend on moist environments, so look in the soft mud along the shores of lakes and ponds for their distinctive tracks. While you may be able to distinguish frog tracks from toad tracks, since they generally move differently, it can be very difficult to identify the species. In drier environments, reptiles outnumber amphibians, but dry terrain does not show prints well, except in sand. Snakes can leave distinctive tracks as they wind their way through mud or sand; foot-less, they make body prints.

Print
Length: to 1.5 in (3.8 cm)

Straddle
1.0–1.5 in (2.5–3.8 cm)

Stride
Hopping: 1.5–5 in (3.8–13 cm)

Size
5.5–6.5 in (14–17 cm)

DARK-EYED JUNCO
Junco hyemalis

This common small bird typifies the many small hopping birds found in this region. Each foot has three forward-pointing toes and one longer toe at the rear. The best prints are left in snow, although in deep snow the toe detail is lost, and the feet may show some dragging between the hops.

A good place to study these types of prints is near a birdfeeder. Watch the birds scurry around as they pick up fallen seeds, then have a look at the prints that they have left behind.

Similar Species: The size of the toes may indicate what kind of bird you are tracking—larger birds have larger footprints. Not all birds are present year-round, so keep in mind the season when tracking. A similar alternating track in open fields is likely from the Horned Lark (*Eremophila alpestris*), a year-round resident of California that prefers to run rather than hop.

Print
Length: to 4.0 in (10 cm)

Straddle
to 4.0 in (10 cm)

Stride
Walking: up to 6.0 in (15 cm)

Size
24 in (61 cm)

COMMON RAVEN
Corvus corax

This legendary bird spends a lot of time strutting around on the ground–confident behavior that perhaps hints at its intelligence. The Common Raven leaves a typical alternating track, with three thick toes pointing forward and one toe pointing backward. When a Common Raven needs greater speed, perhaps in preparing for take-off, it will bound along, leaving irregular pairs of diagonally placed prints with a longer stride between each pair.

Similar Species: Other members of the crow family, all of which spend a lot of time poking around on the ground, have similar tracks with shorter strides. The American Crow (*C. brachyrhyncos*) has prints up to 3 inches (7.6 cm) long. The Black-billed Magpie (*Pica pica*) of the eastern margins of the state has prints up to 2 inches (5.1 cm) long. The Yellow-billed Magpie (*P. nuttalli*) of the Central Valley also has prints up to 2 inches (5.1 cm) long.

Print
Length with rear toe:
    to 1.5 in (3.8 cm)

Straddle
to 2.0 in (5.1 cm)

Stride
0.5–1.5 in (1.3–3.8 cm)

Size
10 in (25 cm)

CALIFORNIA QUAIL
Callipepla californica

    This pretty bird has an unusual crest that is often described as having a teardrop shape. It is widespread in open woodland and in brushy areas and is often found in suburbs, though it usually stays fairly close to a source of water. If you are lucky, in winter you may see a group of up to two hundred birds.

    A California Quail's foot has three forward-pointing toes and a reduced rearward-pointing toe that usually registers only when the quail is walking slowly or standing still, especially in soft mud.

Similar Species: Other quails have similar tracks; for instance, the Mountain Quail (*Oreortyx pictus*), which inhabits the mountainous regions of the state. The Blue Grouse (*Dendragapus obscurus*) also has similar tracks.

Strike
Width to 3.0 ft (91 cm)

Size
22 in (56 cm)

GREAT HORNED OWL
Bubo virginianus

This wide-ranging owl is often seen resting quietly in trees during the day, as it prefers to hunt at night. This owl is an accomplished hunter in snow, and the 'strike' that it leaves can be quite a sight if it registers well. The owl strikes through the snow with its talons, leaving an untidy hole, which is occasionally surrounded by imprints of wing and tail feathers. These feather imprints are made as the owl struggles to take off with possibly heavy prey. The owl is not the most graceful of walkers, preferring to fly away from the scene. If you do find walking tracks, they will look like a larger version of the Burrowing Owl's (p. 132).

You might stumble across this owl's strike and guess that its target could have been a vole scurrying around underneath the snow. If you are a really lucky tracker, you will have been following the surface track of an animal to find that it abruptly ends with this strike mark, where the animal has been seized by an owl.

Similar Species: The Common Raven (p. 126) also leaves strike marks, usually with much sharper feather imprints.

Print
Length: 1.8 inches (4.6 cm)
Straddle
2.5 in (6.4 cm)
Stride
1.0–6.0 inches (2.5–15 cm)
Size
9.5 in (24 cm)

BURROWING OWL

Athene cunicularia

This alert little owl of the plains and open areas spends a lot of time on the ground, bobbing up and down to look out for danger. It inhabits abandoned burrows, such as those made by ground squirrels, often close to burrows still inhabited by these rodents. Many of the Burrowing Owls in southern California are permanent residents, but owls seen in the northeastern counties could be seasonal migrants from further south.

If you are investigating rodent burrows in open areas and you find a burrow entrance with a profusion of tracks that do not have characteristic rodent features, you may have found the unusual tracks of the Burrowing Owl. The print of this owl shows two large, forward-pointing toes with talons, with two additional toes that point off to the side and to the rear. The side and rear toes do not register as well as the front ones, which have more weight on them.

Similar Species: The tracks of other owls are similar, but are unlikely to be found in Burrowing Owl terrain.

Print
Length: 2.0 to 2.5 inches (5.1–6.4 cm)

Straddle
4.0 in (10 cm)

Stride
to 4.0 inches (10 cm)

Size
23 in (58 cm)

MALLARD
Anas platyrhynchos

female

male

This dabbling duck—the male a familiar sight with its striking green head—is common in open areas by lakes and ponds through much of southern California. Its webbed feet leave prints that can often be seen in abundance along the muddy shores of just about any waterbody, including those in urban parks.

The webbed foot of the Mallard has three long toes that all point forward. Though the toes register well, the webbing between them does not always show on the print. The inward-pointing feet give the Mallard a pigeon-toed appearance, which perhaps accounts for its waddling gait, a characteristic for which ducks are known.

Similar Species: The many gulls, and dabblers such as the warm-colored Cinnamon Teal (*A.cyanoptera*) and the elaborately decorated Wood Duck (*Aix sponsa*), are examples of the many waterfowl that have similar prints. Exceptionally large prints of this type are likely from a goose, such as the winter-visiting Canada Goose (*Branta canadensis*), or perhaps the Tundra Swan (*Cygnus columbianus*).

135

Print
Length: 0.8–1.3 in (2.0–3.3 cm)
Straddle
to 1.5 in (3.8 cm)
Stride
Erratic
Size
7.0–8.0 in (18–20 cm)

SPOTTED SANDPIPER
Actitis macularia

The bobbing tail of the Spotted Sandpiper is a common sight on the shores of lakes, rivers and streams, but you will usually find only one of these territorial birds in any given location. Because of its excellent camouflage, likely the first you will see of this bird will be when it flies away, its fluttering wings close to the surface of the water.

As it teeters up and down on the shore, it leaves trails of three-toed prints. Its fourth toe is very small and faces off to one side at an angle. Sandpiper tracks can have an erratic stride.

Similar Species: All sandpipers and plovers, including the common Killdeer (*Charadrius vociferus*), have similar tracks, although there is much diversity in print size.

Print
Length: to 6.5 in (17 cm)

Straddle
8 in (20 cm)

Stride
9.0 in (23 cm)

Size
4.2–4.5 ft (1.3–1.4 m)

GREAT BLUE HERON
Ardea herodias

The refined and graceful image of this large heron symbolizes the precious wetlands in which it patiently hunts for food. Still and statuesque as it waits for a meal to swim by, the Great Blue Heron will walk from time to time, perhaps to find a better hunting location. Look for its large, slender tracks along the banks or mudflats of waterbodies.

Not surprisingly, a bird that lives and hunts with such precision walks in a similar fashion, leaving straight tracks that fall in a nearly straight line. Look for the slender rear toe in the print.

Similar Species: Cranes (*Grus* spp.) have similar habitats and similar prints, but their smaller hind toes do not usually register.

Print
Length: 3.0 in (7.6 cm)
Straddle
Standing: 4.0 in (10 cm)
Stride
(varies with speed) to 12 in (30 cm)
Size
23 in (58 cm)

ROADRUNNER
Geococcyx californianus

   The Roadrunner is a resident of large, open desert
regions and the Central Valley. Like the popular cartoon
character of the same name, the Roadrunner spends most
of its time speeding across the plains on its strong legs.
With its keen eyesight and tough bill, it hunts and eats
insects, small reptiles, rodents and even other birds.
   A Roadrunner's foot has two forward-pointing toes and
two rearward-pointing ones that are much larger: about
3 inches (7.6 cm) in length. The four-pointed star shape of
the print is so distinctive as to be unmistakable, even in
sand. The Roadrunner hardly ever flies; its preference for
staying on the ground and its love for dry areas can result
in a mass of tracks.

Similar Species: Few tracks are like those of this ground-
loving member of the cuckoo family.

toad

frog

Straddle
to 2.5 in (6.4 cm)

Straddle
to 3.0 in (7.6 cm)

TOADS AND FROGS

The best place to look for toad and frog tracks is along the muddy fringes of waterbodies, though toad tracks may be found in drier areas. In general, toads walk and frogs hop, but toads also hop, especially when being hassled by overly enthusiastic naturalists.

Great Basin Spadefoot

Toads leave rather abstract prints as they walk–the heels of the hind feet do not register. In mud, the long toes leave draglines. Most widespread is the Western Toad (*Bufo boreas*), which frequents streams, meadows and woodlands. Spadefoots are toads with a tendency to burrow out of sight in sandy soils–look for a small hill of sand surrounded by toad prints. The Western Spadefoot (*Scaphiopus hammondi*) is a resident of the Central Valley and coastal counties. The Great Basin Spadefoot (*Spea intermontana*) is found along the eastern margins of the state.

When a frog hops, its two small fore prints register in front of the long-toed hind prints. In coastal regions, one of the frogs whose tracks you might find is the large Red-legged Frog (*R. aurora*). The Foothill Yellow-legged Frog (*Rana boylei*) prefers the sunny banks of gravelly streams at lower elevations in the Coast Ranges and Sierra Nevada. An unusually large track is surely from the robust Bullfrog (*R. catesbeiana*), at up to 8 inches (20 cm) in length; this giant, introduced from the east, is spreading its range.

Straddle
to 3.0 inches (7.6 cm)

NEWTS AND SALAMANDERS

*California
Slender
Salamander*

The lack of suitable wet habitats in much of
arid southern California means that there are fewer
species of newts and salamanders than there are further
north. These long, slender lizard-like amphibians live in
moist or wet areas–under logs near waterbodies and in
woodlands. Look for tracks in the soft mud of a woodland
path. In general, the fore print shows four toes, while the
larger hind print shows five (four for the Ensatina). A dragging belly or a thick tail often obscures such detail.

The only newt in this part of the state is the California
Newt (*Taricha torosa*). Found in the evergreen and oak forests of the Coast Ranges and the foothills of the Sierra
Nevada, it reaches a length of nearly 8 inches (20 cm). After
a rain, these newts may leave small trails in the mud as they
emerge from ponds.

The slightly smaller–and highly variable–Ensatina
(*Ensatina eschscholtzi*) is found in the forests of the Coast
Ranges and the Sierra Nevada. The California Slender
Salamander (*Batrachoseps attenuatus*) shares the same range
as the Ensatina, but only as high as the foothills. The striking California Tiger Salamander (*Ambystoma californicus*)–
shiny black, with contrasting pale spots–grows to 8.5 inches
(22 cm) in length. It lives in the foothills of the Sierra
Nevada, across the Central Valley to the coast and south-
wards as far as Santa Barbara County.

LIZARDS

Western
Fence Lizard

Lizards are shaped like salamanders, with similar tracks, but with longer toe marks. However, these reptiles are suited to a variety of dry terrain, rather than moist environments. Therefore, there are many kinds in California.

The attractively marked Banded Gecko (*Coleonyx variegatus*) is common in the southern and eastern desert counties and in the southernmost coastal counties. Never too far from shelter in the rocky and sandy terrain it prefers, it hides by day to avoid the intense heat.

The widespread Western Fence Lizard (*Sceloporus occidentalis*), to a length of over 9 inches (23 cm), is found in many different habitats, except deserts. The Western Whiptail (*Cnemidophorus tigris*), to 12 inches (30 cm) long, can be found anywhere from arid areas to open woodlands, except in the southeastern desert. The larger Southern Alligator Lizard (*Gerrhonotus multicarinatus*), as much as 17 inches (43 cm) in length, is also found in many different habitats–even in wetter areas–throughout the Sierra Nevada foothills and the coastal counties, but not in the San Joaquin Valley.

The most likely skink, to over 9 inches (23 cm) in length, is the Western Skink (*Eumeces skiltonius*). It is found in the Coast Ranges and coastal counties, often near rocks in woodlands and in grassy areas.

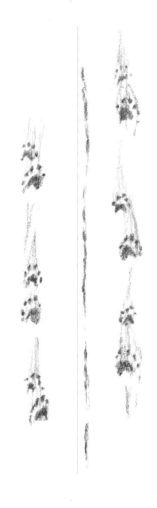

FRESHWATER TURTLES AND TORTOISES

With its large, cumbersome shell and short legs, a turtle leaves a track that is wide relative to the length of its stride—its straddle is about half the length of its body. Longer-legged turtles can raise their shells off the ground, but those with short legs may let theirs drag, as shown in their tracks. The tail may leave a straight dragline in mud. On firmer surfaces, look for distinct claw marks.

Desert Tortoise

The only freshwater turtle indigenous to California is the widespread Western Pond Turtle (*Clemmys marmorata*), which inhabits small lakes or ponds in most western regions of the state. This shy animal will happily slip into the murky depths of the water to avoid detection. It does, however, come out from time to time. To only 7 inches (18 cm) in length, this turtle leaves its distinctive, small tracks along the banks of ponds and rivers and in other moist areas when it comes out to feed, bask in the sunshine or lay eggs.

Slightly larger, to a length of 10 inches (25 cm), the Red-eared Slider (*Chrysemys scripta*) is a popular pet that has escaped or been released in many areas. More abundant near built-up areas, it may outnumber the native turtle.

A turtle-type track in the dry deserts of the southeast is by the Desert Tortoise (*Gopherus agassizii*). A slow, stocky animal to a length of 14 inches (36 cm), it feeds amongst the creosote bushes and cacti. It burrows to keep cool.

149

SNAKES

Common Garter Snake

Snakes thrive in California. The one most frequently encountered is the region-wide Common Garter Snake (*Thamnophis sirtalis*), which is often close to wet or moist areas. Harmless, it can reach 4.3 feet (1.3 m) in length.

Small and secretive, the nocturnal Night Snake (*Hypsiglena torquata*), to only 26 inches (66 cm), inhabits the foothills of the Coast Ranges and the Sierra Nevada. The widespread Racer (*Coluber constrictor*), to 6.4 feet (2.0 m) long, can be encountered anywhere from open fields to wooded hillsides. The Common Kingsnake (*Lampropeltis getulus*), which enjoys habitats throughout the state, can reach a length of 6.8 feet (2.1 m). Sturdy and strong, and found in most of the state, the Pine-gopher Snake (*Pituophis melanoleucus*) can reach up to 8.3 feet (2.5 m) in length.

The most widespread rattlesnake, found anywhere from coast to timberline, is the adaptable Western Rattlesnake (*Crotalus viridis*), which grows to a length of 5.3 feet (1.6 m). From a tracking perspective, the most distinctive rattler, to 33 inches (84 cm) in length, is the swift, agile Sidewinder (*Crotalus cerastes*), which leaves J-shaped marks. Other snake tracks are so similar that identification of the species is next to impossible, though consideration of habitat and range may help. It can be tricky even to establish in which direction the snake traveled.

PATTERNS & PRINTS

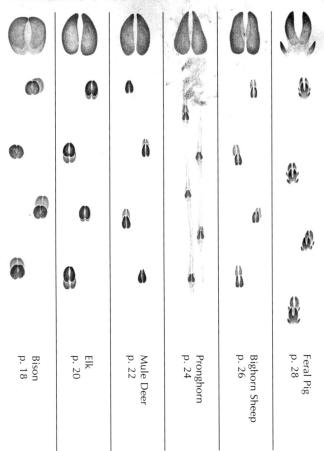

Bison
p. 18

Elk
p. 20

Mule Deer
p. 22

Pronghorn
p. 24

Bighorn Sheep
p. 26

Feral Pig
p. 28

Horse
p. 30

Black Bear
p. 32

Coyote
p. 34

Kit Fox
p. 36

Gray Fox
p. 38

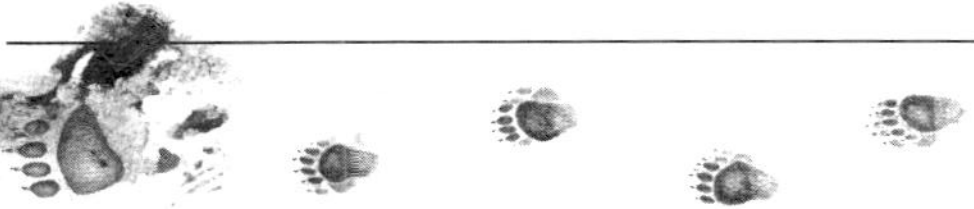

Mountain Lion
p. 40

PATTERNS & PRINTS

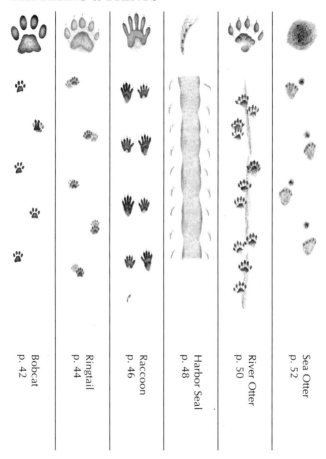

Wolverine
p. 54

Fisher
p. 56

Marten
p. 58

Mink
p. 60

Long-tailed Weasel
p. 62

Short-tailed Weasel
p. 64

PATTERNS & PRINTS

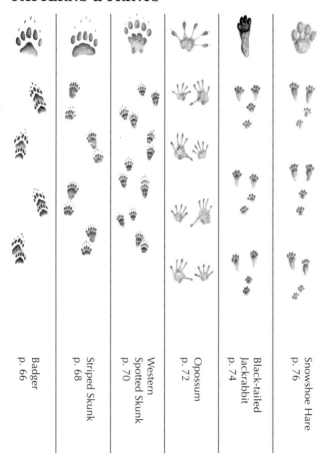

Badger
p. 66

Striped Skunk
p. 68

Western
Spotted Skunk
p. 70

Opossum
p. 72

Black-tailed
Jackrabbit
p. 74

Snowshoe Hare
p. 76

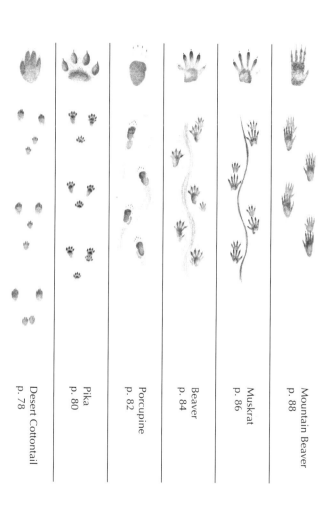

Mountain Beaver
p. 88

Muskrat
p. 86

Beaver
p. 84

Porcupine
p. 82

Pika
p. 80

Desert Cottontail
p. 78

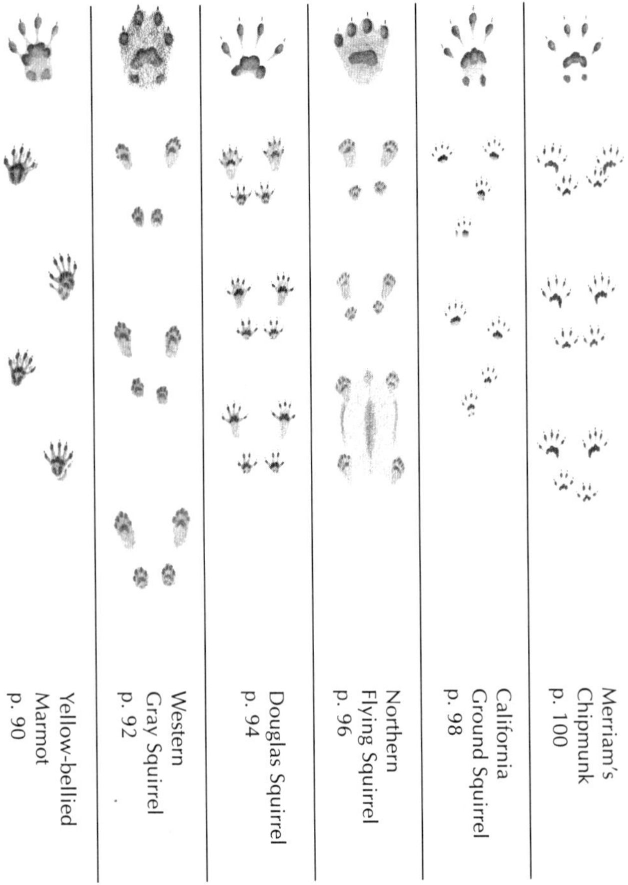

PATTERNS & PRINTS

Merriam's
Chipmunk
p. 100

California
Ground Squirrel
p. 98

Northern
Flying Squirrel
p. 96

Douglas Squirrel
p. 94

Western
Gray Squirrel
p. 92

Yellow-bellied
Marmot
p. 90

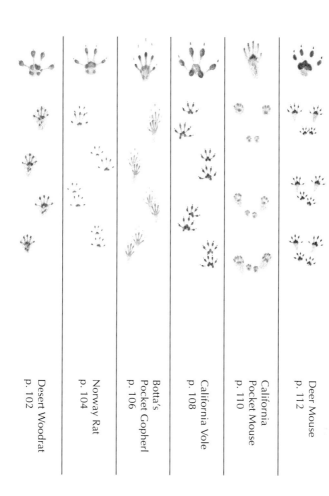

Deer Mouse
p. 112

California
Pocket Mouse
p. 110

California Vole
p. 108

Botta's
Pocket Gopherl
p. 106

Norway Rat
p. 104

Desert Woodrat
p. 102

PATTERNS & PRINTS

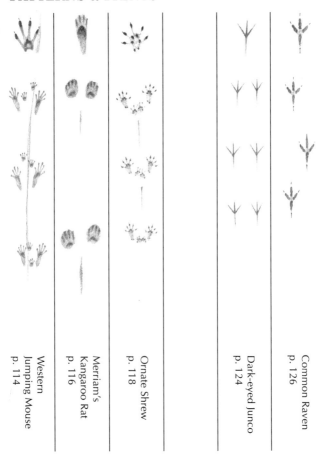

Western Jumping Mouse
p. 114

Merriam's Kangaroo Rat
p. 116

Ornate Shrew
p. 118

Dark-eyed Junco
p. 124

Common Raven
p. 126

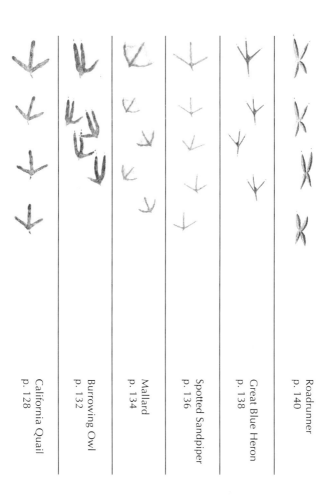

PATTERNS & PRINTS

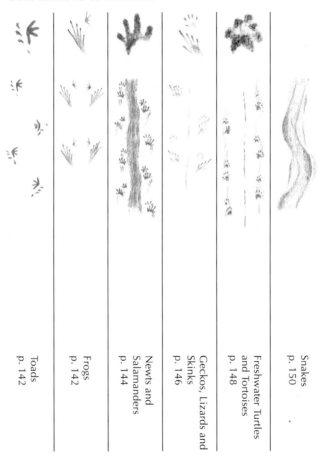

HOOFED PRINTS

Mule Deer   Elk

Bighorn Sheep   Pronghorn   Feral Pig
               Antelope

Bison   Horse

FORE PRINTS

Kit Fox

Gray Fox

Coyote

Bobcat

Mountain Lion

Black Bear

inch cm
0 ⊤ 0

1

2 ⊥ 5

164

FORE PRINTS

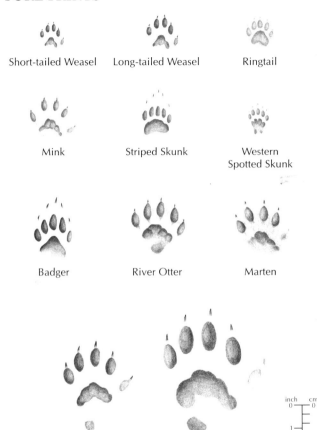

Short-tailed Weasel

Long-tailed Weasel

Ringtail

Mink

Striped Skunk

Western
Spotted Skunk

Badger

River Otter

Marten

Fisher

Wolverine

inch   cm
0   0

1

2   5

HIND PRINTS

Opossum

Raccoon

Muskrat

Mountain
Beaver

Yellow-bellied
Marmot

Desert
Cottontail

Snowshoe
Hare

Black-tailed
Jackrabbit

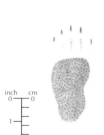

Porcupine

Beaver

Sea Otter

inch cm
0   0

1

2   5

166

HIND PRINTS

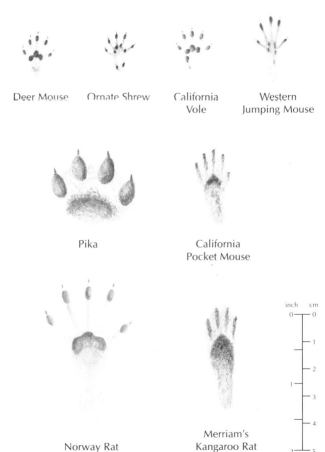

Deer Mouse

Ornate Shrew

California Vole

Western Jumping Mouse

Pika

California Pocket Mouse

Norway Rat

Merriam's Kangaroo Rat

inch   cm
0 — 0

— 1

— 2

1 —
— 3

— 4

2 — 5

HIND PRINTS

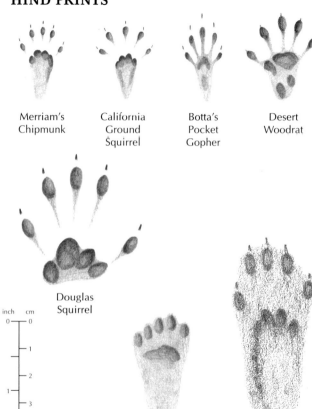

Merriam's
Chipmunk

California
Ground
Squirrel

Botta's
Pocket
Gopher

Desert
Woodrat

Douglas
Squirrel

inch   cm
0   — 0

— 1

— 2

1

— 3

— 4

2 — 5

Northern
Flying Squirrel

Western
Gray Squirrel

BIBLIOGRAPHY

Behler, J.L. and F.W. King. 1979. *Field Guide to North American Reptiles and Amphibians.* National Audubon Society. New York: Alfred A. Knopf, Inc.

Burt, W.H. 1976. *A Field Guide to the Mammals.* Boston: Houghton Mifflin Company.

Farrand, J., Jr. 1995. *Familiar Animal Tracks of North America.* National Audubon Society Pocket Guide. New York: Alfred A. Knopf, Inc.

Forrest, L.R. 1988. *Field Guide to Tracking Animals in Snow.* Harrisburg: Stackpole Books.

Halfpenny, J. 1986. *A Field Guide to Mammal Tracking in North America.* Boulder: Johnson Publishing Company.

Headstrom, R. 1971. *Identifying Animal Tracks.* Toronto: General Publishing Company, Ltd.

Jameson, E.W., Jr. and H.J. Peeters. 1988. *California Mammals.* Berkeley: University of California Press.

Murie, O.J. 1974. *A Field Guide to Animal Tracks.* The Peterson Field Guide Series. Boston: Houghton Mifflin Company.

Rezendes, P. 1992. *Tracking and the Art of Seeing: How to Read Animal Tracks and Sign.* Vermont: Camden House Publishing, Inc.

Stall, C. 1990. *Animal Tracks of Southern California.* Seattle: The Mountaineers.

Stokes, D. and L. Stokes. 1986. *A Guide to Animal Tracking and Behaviour.* Toronto: Little, Brown and Company.

Whitaker, J.O., Jr. 1996. *National Audubon Society Field Guide to North American Mammals.* New York: Alfred A. Knopf, Inc.

INDEX

Page numbers in **boldface** type refer to the primary (illustrated) treatments of animal species and their tracks.

ABOUT THE AUTHOR

Ian Sheldon has lived in South Africa, Singapore, Britain and Canada. Caught collecting caterpillars at the age of three, he has been exposed to the beauty and diversity of nature ever since. He was educated at Cambridge University, England, and the University of Alberta. When he is not in the tropics working on conservation projects or immersing himself in our beautiful wilderness, he is sharing his love for nature. An accomplished artist, naturalist and educator, Ian enjoys communicating passion through the visual arts and the written word, in the hope that he will inspire love and affection for all nature.